FOR KING AND COUNTRY

FOR KING AND COUNTRY

British Airborne Uniforms, Insignia
& Equipment in World War II

1st Airborne Division • 6th Airborne Division
1st Polish Independent Parachute Brigade

Harlan Glenn

Schiffer Military History
Atglen, PA

ACKNOWLEDGMENTS

For undying love and support: my wife Nicole and newborn son George who cries for attention when Daddy's consumed with his work and typing away on the computer.

For support, love and understanding: Mum n' Dad, Sister Vanessa. and last but not in the very least Fred Munferd. As they say man's best friend is loyal until the end. R.I.P. my little friend.

For Inspiration: Bill Costly for reinspiring me to be a writer, and my step uncle Wally Walsh for turning me into an Airborne enthusiast.

For support, time and effort: Nick Hoffman, Mike, Kevin and Nancy of Camera Craft N. Hollywood for their over-the-counter lessons and assistance. Before this I'd little experience taking photographs in a professional manner for publication such as this, so I prayed a lot. John Natland, Neil King and family, Nigel (Pvt. Percy) Horne, Adrian Stevens, James "D" Sharples, Mr. Wybo Boersma Director of the Airborne Museum Oosterbeek The Netherlands, Sean, Dean and Carol Foster, Logistics Museum (UK), The Sikorski Institute, Mr. and Mrs. Jan Lorys, Robert Anzuoni at the 82nd Airborne Museum, Ft. Bragg, Uncle Ken and Aunt Denise of the Croxley Green Family Hospice, Cpl. George Jones C Troop No.4 Cdo. and the Normandy Veterans Association, Harry Workman, Tom Hoare (RIP) 3rd Bn.1st Div., R.J. Anderson C Troop No.4 Cdo., Major J. Dunning (C Troop No.4 Cdo.), Major Gerrard R. Millar, Tish Rayner, Ron Ball (RA) and George, Michael Burgess, Malcolm Johnson, Robert Ehrengruber, Stan Wolcott, Remy Spezzano, Adam Williams, Darren Berry, Nick Clark, the Vander Meij brothers (Polish Bde. recreations), David Gordon, Jez Marren, Gary Archer, Col. H.J. Sweeney for his seal of approval to my effort and a for further enriching my personal inspirations, and Bob Biondi of Schiffer Publishing for giving me a shot. Thanks Mate!

For the loan of materials: John Natland, Michel De Trez, The Summer of '44, Sean and Dean Foster, the family Vander Meij, David and Page Gordon, Andy Birt, Mike Dymkowski, Gary Allen and VERA's 1st Airborne, Fred Miller, Matt Dorn, and Chad Ehler.

Book design by Robert Biondi.

Copyright © 1999 by Harlan Glenn.
Library of Congress Catalog Number: 98-88845.

All rights reserved. No part of this work may be reproduced or used in any forms or by any means – graphic, electronic or mechanical, including photocopying or information storage and retrieval systems – without written permission from the copyright holder.

"Schiffer," "Schiffer Publishing Ltd. & Design," and the "Design of pen and ink well" are registered trademarks of Schiffer Publishing, Ltd.

Printed in China.
ISBN: 0-7643-0794-0

We are interested in hearing from authors with book ideas on related topics.

Published by Schiffer Publishing Ltd. 4880 Lower Valley Road Atglen, PA 19310 Phone: (610) 593-1777 FAX: (610) 593-2002 E-mail: Schifferbk@aol.com. Visit our web site at: www.schifferbooks.com Please write for a free catalog. This book may be purchased from the publisher. Please include $3.95 postage. Try your bookstore first.	In Europe, Schiffer books are distributed by: Bushwood Books 6 Marksbury Road Kew Gardens Surrey TW9 4JF England Phone: 44 (0)181 392-8585 FAX: 44 (0)181 392-9876 E-mail: Bushwd@aol.com. Try your bookstore first.

Contents

Foreword 6
Introduction 8
Dedication 9
Prologue 10

Chapter I	The Parachute Brigade 12
Chapter II	The Airlanding Brigade 29
Chapter III	Corps and Administrative Departments 36
Chapter IV	The Glider Pilot Regiment 58
Chapter V	Issue Clothing 62
Chapter VI	Headgear 89
Chapter VII	Airborne Insignia 97
Chapter VIII	Pattern 37 Webbing 114
Chapter IX	Personal Items of the Airborne 126
Chapter X	Specialist Airborne Equipment 133
Chapter XI	Airborne Vehicles 145
Chapter XII	Airborne Weapons 149
Chapter XIII	'Attached' Troops of the Airborne 167
Chapter XIV	'Tobie Ojczyzno' - The 1st Polish Independent Parachute Brigade 171

Epilogue 188
Bibliography 189
Addendum 190

Foreword

by Col. H.J. Sweeney, MC.

I am honored to be invited by Harlan Glenn to write a foreword to his study of World War II Uniforms, Insignia, and equipments as used by men of the 1st and 6th Airborne Divisions and the Polish Parachute Brigade. As an infantry junior officer I joined the 1st Airborne Division in early 1942 soon after that brigade's change of title and role.

When 1st Airlanding Brigade was divided to form 1st and 6th Airlanding Brigades my battalion became part of the 6th Airlanding Brigade. Consequently I remained with that Brigade which became part of 6th Airborne Division until its disbandment in Palestine in 1948.

Mr. Glenn has produced an accurate compilation and evocation of those early days of airborne warfare when we dropped by parachute out of a cradle attached and through a hole in the floor of a Whitley Bomber. Parachuting was in its infancy and there were many accidents, sometimes fatal.

Mr. Glenn then illustrates by meticulous research how this new form of warfare developed, spawning a whole host of new uniforms, insignia and specialized equipment over the next four years.

It is interesting to note that the British Army had no airborne troops before World War II. The airborne forces grew from a small unit of about 500 men at the end of 1940 to a force of over 35,000 airborne troops by 1944.

He does not forget the gliderborne troops who eventually formed the greater part of the Airborne Divisions and in particular those young glider pilots (many recruited from the RAF) who landed their cumbersome aircraft with great accuracy in spite of being shot at by the enemy and indeed by their own side.

Airborne units of the type illustrated in this book no longer exist. The big divisional formations are quite different. The glider after its short life passed into history. The parachutist now has a limited role and we shall never again see those massed drops, as on the Rhine Crossing, when over 14,000 airborne troops, American, British, and Canadian flew together across the Rhine under the command of General Matthew Ridgway in March of 1945; a massive demonstration of air power and its supremacy. The helicopter has taken over the airborne attack role, and many other changes have taken place. This is why it is so important that someone of Mr. Glenn's ability has taken on the task of recording the history and development of World War II airborne forces whilst there are still a few veterans about who can recall those early days.

Above: Colonel H.J. Sweeney, MC. and his wife Geraldine, in 1942. Note the Colonel's Service Dress tunic with Light Infantry buttons.

MY DAYS WITH THE AIRBORNE FORCES
by H.J. Sweeney

I joined the 2nd Oxford & Bucks Lt. Infantry on 28 February 1942, the very day when Major John Frost attacked the radar post in Brumeval, France with B Company of 2nd Para Bn., the first successful British parachute operation of the war. I brought with me two junior officers and 50 NCO's and men, these were the first reinforcements to 2nd Bn. Oxf & Bucks had since they were designated as an Airlanding Battalion. I was posted to D Company, commanded by Major John Howard who later led the 'coup de main' party which attacked and captured the two bridges over the Caen Canal and River Orne, the first action in the British sector of the Normandy invasion.

I went on a parachuting course in May of 1942 at Ringway, Manchester when we qualified after two balloon descents and five from a Wellington bomber through a hole in the floor.

For the Normandy operation John Howard's boosted company of six platoons and including Royal Engineers flew in six Horsa gliders and had a thirty minutes start on the remainder of 6th Airborne Division to allow us to carry out our 'coup de main' attack before the enemy had been alerted by the activities of other airborne troops. The attack was extremely successful, five out of the six gliders landed accurately, two including my platoon on the River Orne Bridge and three on the Canal Bridge. The capture of these two bridges intact was vital for the protection of the eastern flank of the invasion. The sixth platoon which went astray turned up 24 hours later.

Subsequently I was wounded while on a patrol a month later and evacuated to the UK. After some time in hospital and convalescing, I rejoined my battalion and took part as a captain in 'S' (Support) Company in the Rhine Crossing operation in March of 1945. We had very heavy casualties and both my Company Commander and 2nd in charge were killed, and so I found myself commanding a company for the rest of the war, eventually being promoted to the rank of Major.

After the war in Europe ended, we prepared to go to the Far East but when the atom bomb brought an end to Japanese resistance, we were switched to Palestine where 6th Airborne Division was broken up and we reverted to being an ordinary infantry battalion.

I elected to stay in the Army and made it my career until 1974 when I retired as a full Colonel. I served at Regimental duty in Palestine, Egypt, Cyprus and Malaya/Borneo where I was Commanding Officer of my battalion. I qualified as a Staff Officer and served in that capacity in Berlin, Canada, the War Office and Germany. I graduated from the British Staff College, Camberley and the U.S. Command and Staff College, Fort Leavenworth. I ended my career as a full Colonel in the UK Mission to the UN, New York and finally as Deputy Commandant in the British School of Infantry, Warminster, Wiltshire.

Top: This photo, taken at Ringway in 1942, shows Color Sergeant Major Crewe and Col. Sweeney, who was at the rank of Lieutenant – C.S.M. Crewe was later killed in Normandy, 1944. Of interest are the bungey training helmets and the 'Step-in' gabardine jump smocks which both men are wearing. Note the brass snaps on the Colonel's smock. Once the smock was zipped up, this flap was secured by a series of brass snaps as seen here. It was regulation to wear the P-37 waist belt as shown here while in training, even though no other webbing was worn – this was done to keep a 'smart' look. Also note that, as an officer, the Colonel wears a collared shirt and tie. "P.T.S." is stenciled onto the Colonel's helmet. This probably stands for Parachute Training school.

Above: Field Marshall Montgomery pins the Military Cross on Col. Sweeney (shown here as a Captain), Germany, 1945. Sweeney: "The actual medal was later awarded by His Majesty King George VI. Monty had a very good publicity campaign going as each time he pinned a medal on someone a picture was taken." Of interest are the Captain's pips upon his battledress shoulder strap. These pips look to be embroidered and raised as was the common practice to make them stand out and readily recognizable. The late war white on scarlet shoulder titles is worn, and the AIRBORNE strip has been deleted from underneath the Pegasus flash – you can just see where the AIRBORNE strip was once sewn. The lanyard around Sweeney's shoulder is a Battalion lanyard of Kelly Green which was the color of the Light Infantry. It was very common for the various Battalions to produce and issue their own specially colored lanyards. Montgomery wears the 21st Army Group flash on his sleeve and the rank of Field Marshall upon his shoulder strap. One last note is that the Montgomery is wearing the cap badge of the Parachute Regiment upon his beret in conjunction with this badge of rank. As a morale booster, Monty often wore the badge of the Regiment for which he was visiting.

Introduction

When I became interested in collecting World War II British and Commonwealth militaria I immediately became aware of the severe lack of, and more so non existence of, reference materials. It was a case of learn as you go from what little sources you could. Living in America there are no British Regimental museums that I or others could visit. Here you're lucky to see a British helmet! It became necessary for me to travel to the UK and Europe to do research and interview veterans. I wanted a book like those available to collectors of German militaria which I could take with me to a show and with confidence know what I was looking at. Through experience I learned not to trust some militaria dealers as they were simply ignorant in the British and Commonwealth area and would rather sell their white elephants than truthfully tell you what is right for your needs. This is something that new collectors experience and maybe it's good to get 'burned' at least once so the next time they'll do their homework and not have to ask.

This book is a broad based effort, and hopefully will serve as a guide and best friend for those new to the hobby, as well as a detailed reference and intermediate collector and reenactor. Fu- ture volumes will continue with the Airborne theme and also cover the line infantry- man. I have tried not to over- whelm the reader with the ol', "On the first day of the third month of the 7th hour ... the Denison smock was created", but tried to lay a foundation to start with.

My main objective was to discuss as many variations of clothing, insignia, and equipment as used by the World War II Airborne soldier. I do however suggest that you observe this rule of thumb in your collecting travels and that is there were numerous variations of issue and private purchase clothing, insignia, and equipment manufactured during the war. Keep an open mind and try not to formulate your opinions too sternly.

This first volume covers the British and Polish Airborne soldier. Also featured are the 'Attached' liaison troops of the Air- borne Divisions. A later volume will continue with the British airborne theme covering and expanding upon items not dealt with here. Specially featured will be the Canadian Parachute Battalion of the 6th Division.

Some of the items in this book are unobtainable and or very close to it. Do not expect to own many of these items. Should you come across one, you should rejoice, as you are truly one of the few! Have a good read and I hope you enjoy this effort.

Harlan Glenn

Above: A wounded rifleman covers a Bren Gunner as he moves to a better position. Notice the rifleman's fixed bayonet – this was often done in close combat, especially during the Arnhem battle. Both troopers wear full kit order.

Dedication

As Pvt. E.G. Doorland of D Coy (Company) 3rd Parachute Battalion said to me on the front steps of the Hartenstein Museum in 1995, "I was just an ordinary soldier." This book is dedicated to 'ordinary' soldiers like Pvt. Doorland who gave their all for King and Country and for their mates! Words cannot express in all the deepest and most sincere appreciation my thoughts and feelings for these gentlemen. This volume is specially dedicated to the 1st Polish Parachute Brigade who are without a doubt the unsung heroes against adversity.

Prologue

In The Beginning ...

"We ought to have a corps of at least five thousand paratroops," wrote Prime Minister Winston Churchill to his Chief of Staff on June 22nd 1940. That letter is what started it all. It was prompted by the success of the German Airborne in the first months of the war. Churchill wanted to strike back and the Airborne were one of the swords with which he planned to do so.

Training began on July 8th 1940 and by July 21st 1940 the first trainees began their jumps with American made static chutes. Training was suspended July 25th when a man's rigging lines became twisted and he fell to his death. Training resumed again on August 8th with the introduction of the first British made static chutes. These chutes were manufactured by the GQ Parachute Company and designed by an ex-Grenadier Guards Officer Raymond Quilter. Most of the Jump Instructors were Army NCOs since the RAF did not have instructors available until later in the war as they were busy training their own pilots and crewmen of the RAF.

No.2 Commando provided the first guinea pigs/pupils for experimental parachute training. It was a Driver Evans who was attached to No.2 Commando who fell to his death. The first jumps were made by standing in the open tail of a Whitley bomber and allowing the slip stream to pull them off. The man would then open his RAF style chute via a rip cord. When the pupils converted to the American made static chutes they jumped through a hole in the floor of the aircraft. On the 15th of August 1941 No.2 Commando was bestowed the honor of being the very first Parachute Battalion, and so was born the 1st Parachute Battalion. This was the beginning of Britain's newest and proudest regiment – the Parachute Regiment.

Above: A Para at the very beginnings of the Parachute Regiment. This is the first pattern 'step in' smock, designed after the German 1st pattern jump smock – this British variant saw limited combat, later being replaced by the Denison smock for the landings in North Africa. The 1st pattern smock was made of an olive denim material like that used for the denim overalls. Worsted rank was worn with jump wings upon the shoulder. A Field Service cap is worn here with the soldier's parent regimental cap badge – as the maroon beret had not yet been introduced (just prior to the jump into North Africa). Thompson SMG's were used until the Sten SMG was introduced and adopted in 1942. As it was heavy, the Thompson was more a weapon for the line regiments and the Commando, whereas the light weight Sten could be broken down into sections making it easier to stuff under one's jump harness. Issue ammo boots and P-37 anklets complete the kit.

Those Who Served

Written in the veterans' own words, these vignettes are included to offer the reader an insight into a particular unit. These stories will be featured throughout each volume.

Wally Walsh No.2 Cdo., 11th Special Air Service, 1st Parachute Battalion 1st Airborne Division

After escaping through Dunkirk (Wally went over with the British Expeditionary Force in 1940 - Author), I was assigned to a temporary division which was made up of escapees. In this division were men from all sorts, all grouped together. Some time later, a notice made its rounds asking for volunteers for a Special Service Battalion. This was the beginning of the Commando.

I signed up and off I went to Commando school. I changed our name and we were called the 11th SAS. And was posted to No.2 Commando, which later they then later they changed our name again to 1st Parachute Battalion. I remember my first jump. I was scared but I told myself that I had to do it and I knew I would. So when I got to the door, I told myself well this plane can crash and I will die. If I jump and my chute doesn't open, I will die. So, well, here goes nothin'! And out I jumped and once my chute opened it was the greatest feelin' I've ever had. Wonderful! You tell yourself when you do your jump and you'll be ok! Just say ... Here goes nothin'!

I stole my Lieutenant's Humber Snipe once. I wanted to see my mother and sister as it had been a long time since I had last seen them – we were very close, they were all I had. I asked for a pass, but it wasn't granted so I went anyway. They all knew I would. I took his Humber Sniper and spent the weekend with my Mother and Sister. I knew I better get back as they would have the Red Caps (Military Police) after me, and they had. I left just in time. I drove the Lieutenant's Snipe to a village far off from the camp. I urinated on the steering wheel so as to burn off any finger prints, they taught us that ya know. I then climbed over the fence and snuck into camp. The Sergeant of the Guard said that I better go with him as that I was to be arrested for stealing the Lieutenant's transport. I knew they couldn't prove it as I'd removed my prints. The Lieutenant was mad as a Hatter'! He knew I did it but he couldn't prove it. They put me in the guard house for leaving my post but that was all they did to me as I was a good soldier and they knew it. I did my job and I did it well.

My first combat jump was into North Africa, then Sicily, I missed the Italian Toranto landin'. Good thing as I lost some friends on that one. It was terrible, their boat hit a mine and they were killed (this was a seaborne landing for the Airborne). Next came Arnhem! We knew where we were going (Holland) as we were training on flat land much like it. We knew! We knew! We wanted to go so as to not be left out as we had missed the Normandy jump. We were afraid we'd be left out and miss the end of the war without another jump under our belts.

I was a sniper, I used a Bren and shot single shot was about 200 yards. As the days went on and things through the church tower. The closest I got to the bridge were getting bad, we were ordered to get out. I was stripped down and in the water (Rhine river) about to swim across when I saw a wounded Para crouched over, halfway in and out of the water. I couldn't leave him like that. I took my dressing out of my trouser pocket. This was breaking King's Regulations but I couldn't get to his! I began wrapping the bandage round his head, one of his eye balls was hanging out of it's socket, when I felt a sharp hot pain in my hand. I had been shot! A German sniper hidden in the water had got me. We were taken to St. Elizabeth's Hospital I never saw the wounded Para again. I did hear later that he had been shot in both legs but I don't see how he could have tried to escape. He wasn't going anywhere. I don't know why they'd shoot him in the legs? I spent the rest of the war in a prisoner of war camp.

Top: Pvt. Walt "Wally" Walsh prior to the war. Here he wears his Service Dress uniform as issued to other ranks prior to the introduction of Battledress. Of interest is the high collar (for other ranks) SD tunic – officer's SD tunics featured an open collar worn in conjunction with their collared shirt and tie.

Above: Wally and the author during the 1995 50th VE-Day celebrations. Here Wally wears his Regimental blazer and current issue beret. This was a very enjoyable evening, as I walked into the room and he put his arm around me and said, "The battle went like this ..."

Chapter I

The Parachute Brigade

Originally each Parachute Division was to have two Parachute Brigades – the 1st Airborne Division had the 1st, 2nd, and 4th Parachute Brigade; the 6th Airborne Division had the 3rd and 5th Parachute Brigades. After the Sicilian Invasion the 2nd Parachute Brigade remained to fight in Italy. It was then redesigned the 2nd Independent Parachute Group. The 1st Airborne Division went back to England to refit as its ranks were thin from losses in North Africa and Sicily. The 2nd Independent Parachute Group later took part in Operation Anvil the invasion of Southern France. From this point each Division had two Parachute Brigades and one Air Landing Brigade.

Each Parachute Brigade consisted of a Brigade HQ, and three Parachute Battalions. The 1st Brigade consisted of the 1st, 2nd, and 3rd Parachute Battalions. The 2nd Parachute Brigade consisted of the 4th, 5th, and 6th Parachute Battalions. The 3rd Parachute Brigade consisted of the 8th, 9th and 1st Canadian Parachute Battalions. The 4th Parachute Brigade consisted of the 10th, 11th, and 156th Parachute battalions. And the 5th Parachute Brigade consisted of the 7th, 12th, and 13th Parachute Battalions.

Each Parachute Battalion had its own parachute engineer squadron and parachute field ambulance. A parachute battalion was made up of five hundred and fifty men. Each battalion had its own battalion headquarters, headquarters company, machine gun platoon (four Vickers machine guns), mortar platoon (six 3" mortars), and three rifle companies. Each rifle company had five officers and one hundred and twenty men. Broken down into a company headquarters, and three platoons of thirty-six men.

One didn't simply join the Parachute Regiment, you had to fill out an application to join. If your Commanding Officer approved and signed it, your application was then sent on, and an interview would be granted. If the soldier passed the interview, he was then sent to a training camp. If he passed this, then he took the jump course. Here he had to complete seven jumps, two from a balloon, and five from an aircraft. Should a soldier not pass his training and graduate from the jump course and training camp, he would be 'RTU'd (returned to unit, this being his parent regiment). The training was hard, and at times brutal, this came in handy when they were fighting for their lives in desperate combat situations.

A Para in North Africa. Shown here is a first pattern Denison smock worn with issue Parachutist trousers and the newly issued maroon beret, which became the namesake of the British Airborne forces. Note the plastic economy Parachute regimental cap badge. Also newly issued was the No.4 rifle, first introduced for Operation Torch. Also seen is an issue camouflaged face veil – this could be dunked into water and worn around the neck to keep the soldier cool in the desert heat, or used as a sniper's veil.

Chapter One: The Parachute Brigade

Paras break for a smoke. From left to right, an officer offers boiled sweets (hard candy) to his men. Notice the body armour, which was only issued and worn for Arnhem in limited number. Note the beret tucked through the smock's shoulder strap. Rank 'slip ons' are worn over the smock's shoulder straps. Plastic economy pips are worn here. The soldier in the middle has a small bicycle lamp hanging from his pocket, and the Para on the right wears a shell dressing from his shoulder strap, commonly done for the Arnhem battle. Also note the bayonet – this variant was not commonly issued. All wear the late-war web chin strap helmets, which were mainly issued to the Poles and Canadians, as most of the British wore the leather chin strap helmets of earlier issue.

The Requirements to become a member of the Parachute Regiment

The requirements were as follows (taken from "War" Nov. 28th 1942):

1. His color vision must be extremely good for red and green. These were the colors of the lights in the aircraft and also the color of the supply chutes.
2. A man must be under 6' 2" and under 13 1/2 stone (about 182 pounds) because oversized men were more inclined to get hurt upon landing from all that weight crashing down).
3. Eye sight must be good and hearing good, there are also dental standards. A careful investigation was made into any tendency to come down with bronchitis, etc. and or a history of broken bones was taken into consideration.

If a man got this far there was a 50% chance he would be accepted. At the interview it was up to the officer in charge as to what questions were to be asked. Parachute pay was an extra two shillings for other ranks and four extra shillings for an officer. This served as an added incentive for many a soldier to join.

A typical day went something like this: bugle call at 6am and you were out of your bed by the time the last note sounded. Some fifty yards away from the hut (barracks) was the wash house with its sixty basins and more than sixty men will want to wash and shave before breakfast. Some ingenious individuals will get up maybe ten minutes before six to get a head start and therefore perform their toilet

Here the King (in great coat) visits the Paras on a training course. Note the aircraft fuselage in the background – men could climb the ladder, and then jump though a hole in the floor landing below in the sand bag pit. To the right are two Mortarmen – note the 3" bombs and bomb tubes which were made of heavy duty cardboard with metal caps, and web straps to secure and carry the tubes. Also note the padding and drop container for the bomb tubes. Lieutenant General F.A.M. "Boy" Browning (far left) Commander of the Airborne Forces, and Major General R. E. Urquhart (to Browning's right) Commander of 1st Airborne.

13

duties in comfort. At ten past six the basins will be in use by men under great pressure by those desperately wanting to do the same, wash up, pass inspection and eat! After washing up their beds had to be made. The 'biscuit' (mattress) and sheets are folded and stacked as per regulation. The huts are then swept, dusted and mopped. And in some 20 minutes the hut is immaculate. The Company officer and Sergeant Major will then inspect the hut and the men. Those who do not pass will be assigned extra unpopular duties. The call to the cookhouse is called and the men eagerly go off with their mess tins, utensils and tea mugs to breakfast.

A soldier received three meals, breakfast at seven, dinner and eleven, and tea at five. Breakfast might consist of bacon, bread, margarine, porridge and tea. Sausage might substitute for bacon. For dinner there might be a meat, potatoes, greens or beans and "duff" (a boiled rice dish, pudding or fruit tart). Tea may have included eggs and chips (French fries), a meat pie or pastry (meat roll), bread and margarine, jam, and tea.

The day's itinerary consisted of the following: P.T. (physical training) which included bending, stretching, and sports, unarmed combat, kit inspection, and weapon's revision which was the study of the various weapons. This included the rifle, Bren, Sten, grenade, each weapon in detail, and consisted of dismantling and trouble shooting. All basic weapons would be covered and absorbed so as under the stresses of combat the soldier could carry on with his duties. As stated in their weapons manuals the sole object of weapons training is to teach all ranks the most efficient way of handling their weapons in order to kill the enemy.

A day might also include lectures on battle drill (field craft), battle drill practice, first aid, container drill (drop containers), gas test (alert), and the Padre's hour that was designed to stimulate an interest in religion by providing the soldier an opportunity to raise and discuss problems that he might be incurring.

As the Divisions were rapidly growing in size, men were needed to fill out the ranks, but volunteers were not coming in as needed. By June 1st 1943, perks and increases in pay were allocated to encourage recruiting. And in July 1943 the minimum age for volunteers was lowered to 18 1/2 years of age.

Pay as follows:

1. Additional pay for Paratroops and Glider Pilots for any time while in captivity (P.O.W. [Prisoner of War]).
2. Additional pay for those Paratroops and Glider Pilots for up to 91 days in hospital, who had received wounds attributed to their duties.
3. Airborne pay for Glider and Airborne troops in a non-parachute role for those in hospital for up to 91 days suffering from wounds attributed to their duties.
4. Acting rank of Glider Pilots continued for both POWs and while in hospital.

The following Battalions were formed directly from recruiting drives:

2nd Parachute Battalion, 3rd Parachute Battalion and the 4th Parachute Battalion.
The 1st Parachute Battalion was as mentioned formed from No.2 Commando/ 11th SAS. Their ranks were filled and brought up to strength by new recruits from such 'Recruiting Drives'.

Last minute adjustments are made prior to enplaning. Note that no cap badges are worn on the berets. Also note how differently the pistol lanyards are worn, one around the neck, and another down the front, around the shoulder. The Para to the right, with no beret, is wearing his entrenching tool in a very unusual manner, as most wore them on the middle back. Interesting is how each man has stuffed his pockets with bits of kit, making good use of the extra space. This photo suggests that it is 1944-1945, due to the equipment and clothing worn. The kneeling man wears the third pattern web chin strap jump helmet. As a last note, the scrim has been cut in rather large pieces, as seen on the far left Para's helmet.

This Para in the woods of Oosterbeek is on patrol, as he is in light order (meaning no webbing). He wears a .303 ammunition bandoleer and a toggle rope. The helmet is the all steel leather chin strap 1943 pattern. Once the men had been in the field, it was common for them to drop their packs and leave their equipment while going out on a 'Recce' (patrol). This gave them added mobility.

Chapter One: The Parachute Brigade

Redesignations

It was thought that *esprit de corps* would be higher amongst a line unit who had trained and served together to go through the parachute selection course. Many of the Line soldiers were found physically or psychologically unfit for the rigorous training of the Airborne, this was discontinued and recruitment was kept to an all volunteer basis.

The following Battalions were selected from their line regiments and put through the parachute selection course as whole units. Those who passed the course became the founding members of these Parachute Battalions:

7th Battalion Queens Own Cameron Highlanders formed the 5th Scottish Parachute Battalion.
The 10th Battalion Royal Welsh Fusiliers became the 6th Royal Welsh Parachute Battalion.
The 10th Battalion Somerset Light Infantry became the 7th Parachute Battalion.
The 13th Battalion Royal Warwickshire Regiment became the 8th Parachute Battalion.
The 10th Battalion Essex Regiment became the 9th Parachute Battalion.
Volunteers serving with British Infantry Battalions serving in India formed the 10th Parachute Battalion.
The 10th Battalion Green Howards became the 12th Parachute Battalion.
The 2nd/4th Battalion South Lancashire became the 13th Parachute Battalion.
Volunteers serving with British Infantry Battalions serving in India formed the 151st Parachute Battalion. This was redesignated 156th Parachute Battalion.

Each Parachute Brigade consisted of a brigade HQ, and three Parachute Battalions. A parachute battalion was made up of five hundred and fifty of all ranks. Each battalion had its own battalion headquarters, support company which consisted of a machine gun platoon with four Vickers machine guns, a mortar platoon with six 3" mortars, and three rifle companies consisting of five officers and one hundred and twenty other ranks. Each Company had its own broken company headquarters, and three platoons of thirty-six men. Each Brigade also had attached to it a Parachute Engineer Squadron and Parachute Field Ambulance.

The Rifleman

A Rifleman was issued with and carried the following for a typical operation:

One light assault respirator (MKIII), attached to belt.
Basic pouches, each with two Bren gun Magazines or two 2" mortar bombs, attached to a belt and shoulder braces.
One spike bayonet and scabbard held in 'frog' on belt.
One bandoleer with fifty rounds of rifle ammunition.
Two hand grenades carried either in knee pocket of Battledress or in tunic pockets.
One fighting knife
One toggle rope, for climbing, slung round shoulders.
One haversack.
One water bottle and carrier.
One Entrenching tool and carrier.

Taking a 'bead' on a wrestless "Jerry", this Rifleman waits for the shot. Note the angola wool lined collar of his smock and the riveted leather chin strap. Chin straps were both riveted and sewn, some examples seen by the author have been both sewn and riveted. The Rifleman wears the late war 1943-44 helmet net, as the Lance Corporal wears the larger mesh net of earlier issue. Of interest is the chamois lined chin cup of the LCpl.'s helmet chin strap.

Spotting for the Rifleman, the LCpl. directs him for the perfect shot. Note that no rank is worn on the L.Cpl.'s left sleeve. It was common practice among the Airborne to wear only rank chevrons on the right sleeve.

Left: Here a cautious Rifleman prepares to cross a small country lane, careful not to attract a sniper's bullet – snipers were rampant in Arnhem, on both sides. Of interest is the Airborne trousers bellows pocket. Here you can see how it folds and lays upon the trouser leg as it is not sewn corner to corner, rather sewn underneath leaving a 1 1/2" edged flap.

Left: Front view of a Rifleman, 1944-1945. Note the wrist tabs with added hose tops. The hose tops were not regulation and/or issue, but rather a private practice done by the individual. This was done more so after the war than during as there is little photographic evidence to support this as a wartime practice. Also note that all webbing is properly blancoed as per Army regulation. Right: Back view of the Rifleman. Seen here is the P-37 sleeve canteen carrier, known as the 'Airborne' carrier – it is simply another variant rather than a special 'Airborne' item. The 'Ape tail' of the smock is held in place and secured to the back of the garment. Note how the toggle rope is positioned and secured, around the shoulders and then behind the back, and is done by the book. The ground sheet is carried as by regulation in the haversack. What stands out are the polished brass buckles on the waist belt. Though the webbing was blancoed for camouflaged, the brass has been highly polished.

The Sten Gunner

The Sten Gunner was issued with and carried the following for a typical operation:

One light assault respirator, attached to belt.
One Sten gun.
One Sten bandoleer containing seven magazines of twenty-eight rounds each. Note that a Sten mag will hold 32 rounds, yet it is common military practice to under load magazines to prevent jamming. As the spring tension is so great, that if a mag is filled to capacity, the first round fired will jam.
MK III basic pouches. These could hold either two Bren gun Magazines, or two 2" mortar bombs in each pouch. If the Sten bandoleer was not in use, the soldier would carry his Sten magazines in these pouches. MK III pouches were made 1" longer than the MK II pouches in order to accommodate the Sten magazine which could not be snapped closed in the MK II pouch. Either pouches were then attached to the belt and shoulder braces.
One Sten magazine in pocket.
Two hand grenades.
One fighting knife.
One toggle rope.
One haversack.
One water bottle and carrier.
One Entrenching tool and carrier.

Chapter One: The Parachute Brigade

Left: Front view of a Sten Gunner, 1944-45. This soldier's smock has been modified, as seen by the full zip conversion. At the crotch, the Sten sling touches the smock – that is the full length zipper – and was an officer's feature as few other ranks did this. He is armed with a MKV Sten which was commonly issued from Normandy until the end of the war. Attached to his leg is the Tanker Holster. Right: Sten Gunner, back view. Shown here is another way of carrying the toggle rope – this was common among the Glider Pilots. The cord dangling beneath the canteen is the Webley revolver lanyard. Note the leather tabs on the P-37 gaiters – this is a second variant, as others were of an all web construction.

Left: This is the MK III Basic pouch. It was taller/deeper than the MKII in order to accommodate the Sten magazines. Here a Sten Gunner prepares to load a new magazine. Note the brass hook just above the pouch which is how the haversack was secured to the main webbing frame via the L-straps (shoulder straps), which were attached to the haversack, and then hooked through the keeper of the basic pouch. Right: This soldier wears the Austerity Pattern trousers. Parachutist trousers were not issued to all Paratroops, therefore there is no special pocket for the dagger and it must be worn on the belt. Here it is carried in the P-37 bayonet frog. Note how the loop of the toggle rope is secured on the entrenching tool handle.

17

For King and Country: British Airborne Uniforms, Insignia & Equipment in World War II

The Bren Gunner

The Bren Gunner was issued with and carried the following for a typical operation:

One light assault respirator, attached to belt.
One revolver carried at left side.
One .38 pistol ammo pouch with revolver ammunition attached to belt.
Basic pouches, each with two Bren gun Magazines, attached to belt and shoulder braces.
Two hand grenades.
One fighting knife.
One toggle rope.
One haversack.
One water bottle and carrier.
One Entrenching tool and carrier.

A Bren Gunner moves cautiously to a forward position. The Bren could be carried as seen here, or by its top handle above the barrel. There was also a special sling for the Bren which allowed the wearer to take some of the burden off his arms. Behind him is a stalled jeep probably belonging to the Air Landing Brigade. Note the angola lined collar of his Denison.

Below: A fresh magazine is in and the 2nd team's gunner has made it safely to cover. Now is a good time to change the barrel, before "Jerry" makes another counter attack.

Chapter One: The Parachute Brigade

A Bren team covers another as it crosses a deadly lane. Note that the assistant gunner is poised ready with another Bren barrel – each team carried a spare. After so many rounds were shot the barrel was to be changed – this prevented it warping from the heat of overuse. Atop the sand bags is a metal ammunition case which holds ten extra Bren magazines.

Above: The 2nd team's assistant gunner has been hit while crossing. Now the 2nd gunner races to the other side in hopes of not being hit himself. The first teams assistant gunner is about to change to a fresh magazine as the one in the gun is about out. Of note is the captured MP-40 of the 2nd team's assistant gunner. The Airborne had to appropriate weapons, ammunition, and supplies as theirs ran out – especially in Arnhem where there was no true re-supply. Many Paras were seen to be using German K-98 rifles and MG-34s.

Above right: The 2nd Gunner gives it his all to make the other side as the first team keeps up the fire. The 1st team's assistant gunner is concerned as it is well past time to change the barrel. The gun will be no good with a warped barrel, yet until the 2nd gun is up and running, there is nothing he can do.

Right: Now the magazine must be changed as the gunner has spent the other. The crossing gunner must hurl himself to cover to avoid being hit.

For King and Country: British Airborne Uniforms, Insignia & Equipment in World War II

The DZ

Pronounced by the British as 'D-zed' for Drop Zone, meaning ... where those who parachute are 'supposed' to land. Here you'll see a paratrooper who has just hit the ground. His first actions being to collapse the chute, get his equipment sorted (assembled) and on and then head to the RV (rendezvous point), which has been predetermined prior to leaving the aerodrome.

Below: An Officer just touches down as a Bren Gunner gathers himself and heads on the to the rendezvous point. The officer has just collapsed his chute and must now get out of his harness and then get his personal equipment on and make for the RV point himself. The Bren Gunner is wearing his equipment underneath his denim over smock (Jacket Parachutist).

Left: As this officer hits his Quick Release box his harness will unlock and he can then easily remove the harness. This Quick Release box was an innovative piece of equipment and saved many lives – American Airborne did not have one and some drowned in the bogs of Normandy because of it. Note this is a post-war parachute – this chute is identical, but for the color of the harness webbing which during wartime was white. The open flaps that once held the packed chute in place, which now are open, flap back and fourth in the breeze. Of interest are is the fashion in which the soldiers weapon is carried. This was common practice for carrying the Sten Guns as rifles and light machine guns had their own padded valise. The olive denim garment is an oversmock, officially called the Jacket Parachutist, and fits over the Denison and the soldiers webbing. The haversack cannot be worn on the man's back as the bulk of the parachute will not allow. The haversack is carried inside the bottom front of the oversmock, secured by the harness, and pressed against the man's body. Between the mans legs are the securing ape tail for the Jacket Parachutist. Right: Here the Quick release box is open and hangs to the side. The officer lets his weapon drop from the harness as instructed during training: "Get out of your harness, get your weapon ready, and then get your equipment on."

Chapter One: The Parachute Brigade

Left: Note how the Sten MK V has been set in a locked position, thus the fore grip, magazine holder and trigger housing are all even, making it easier to carry in the harness. Once out the gun will be set and locked into position for action and use. Note also the twin grenade pockets of the Jacket Parachutist. Should a Para drop into an unfriendly situation, using grenades, he could quickly make a defensive diversion and escape. Right: The Sten is pulled out to be later assembled. Note that the Para has a loaded magazine with his weapon – if needed to be could go directly into action to defend himself. Some Paras carried their weapons in a padded valise or leg bag, so should a man come under fire he would have to take his weapon out of the bag and then get out of his over smock to get at his ammunition. Therefore many smart Sten Gunners carried a magazine at the ready in their harness.

Left: The Para will now step through the harness, freeing him. Note the long zipper of the Jacket Parachutist. This very zipper was what was used in the full zip Denison conversions. These over smocks would remain behind on the drop field where the R.A.S.C. might come and collect them for another operation if they had the time, and if it was safe to do so. Right: The No.36 was the standard bomb (grenade) carried by the British and Commonwealth soldier.

21

For King and Country: British Airborne Uniforms, Insignia & Equipment in World War II

Left: He will now set down the main body of the weapon. Note that movement was very difficult as the harness was very tight. The Paras were very well trained to be able to hit the ground, get out of their harness, and assemble their weapon. Right: The harness is off, the Jacket parachutist is unzipped, and he now unsnaps the securing crotch/tail piece. As can be seen the Jacket Parachutist has three securing positions for the comfort of the wearer. They were sized with a series of four numbered labels which were sewn to the inside of each smock.

Left: As the Jacket Parachutist has been unsnapped the tail hangs down and now the officer can resume unzipping the smock. The zipper goes the length of the garment and the tail covers about a foot of it. Right: The haversack drops out. Also seen is the body armour underneath the over smock. The soldier is completely dressed and equipped (the exception being those who employed leg bags, where a Paras entire kit could be held) underneath the Jacket Parachutist – notice that there is no lining as it is a simple garment. Just under the smock, atop the officers shoulder is the tip of the rank slip on.

Chapter One: The Parachute Brigade

Left: The size label of the over smock can be seen at the far upper left. Note the zipper pull tab dangling beneath the zipper to the right corner of the Jacket Parachutist. This pull tab was to enable the zipper to be opened and closed with ease. Beneath the Officers feet is the haversack – on its back you can see the L-straps (shoulder straps). To the officer's right shoulder you can see his rank – plastic economy pips which are attached to a khaki twill slip on sleeve. This sleeve is fitted over the shoulder strap of the Denison smock. The piece of twine/string around the neck is a clasp knife lanyard. Each man was equipped with this personal pocket knife which was attached to a lanyard to avoid losing it in combat. Right: Around the officer's neck is the camouflaged face veil. Issued to each man it functioned as a scarf and camouflaged sniper's veil. It could also be thrown over the helmet and head, defusing any hard edges and blending the wearer's profile into the surroundings. Note that the helmet net is of an early issue. The 1st Division would retain older issued equipment, and the 6th more recent issue. (Note the 6th did not see combat until June 6, 1944.).

Left: The Jacket Parachutist is tossed aside as the officer prepares to assemble his weapon. Note the long tail of the smock as it dangles limply. Right: Note how the bottom half of the body armour bends like a hinged door. One can see a better view of the rank slip on from this angle.

23

For King and Country: British Airborne Uniforms, Insignia & Equipment in World War II

Left: Now the gun assembly drill begins ... remember that this all takes about a minute and a half, from the time he hits his quick release box to the minute he walks away. Right: The stock is fitted into the locking lug of the main body. Note that the magazine holder is still in the down position. Also note the small brass buckle on his left shoulder brace, and the top open slit which is for the lipped hook of the L-strap (shoulder strap). This secures the haversack while the soldier is moving.

Left: Of interest in this photo are at left the brass buckles that attach the haversack to the L-straps. The Sten Sling hangs loose until hooked and secured at the barrel end. The Stock is slotted and will then be turned and locked into position. Also note that the body armour can be adjusted to the wearer's size. Right: The stock has been locked into position. Note the brass butt plate and latch cover. Inside are the Sten cleaning tools, the issue pull through and cotton pads, and lastly an oiler (oil bottle).

Chapter One: The Parachute Brigade

Left: The Officer grabs his loaded magazine, twists and secures the barrel into its operating position, and then prepares to load the weapon. Right: The magazine is loaded, the bolt will then be locked and put on safety.

Left: The Ape tail is secured to the back of the smock and now he begins to put on his haversack. Right: Note the well polished brass buckles on his webbing. This officer's batman (personal valet, a position of tradition in the British Army) must be a very busy man. Also note that the smock's shoulder straps are buttoned over the web braces. The haversacks straps will be worn over the smock's shoulder straps as per regulation. The one notable exception to this is when the call of nature arrives. When not simply remedied by unbuttoning his trouser fly, he will have to take off everything else, as he is wearing trouser braces and the trousers cannot be simply dropped. Think about it. If in a hurry he'd have to make a mad dash of it!

25

For King and Country: British Airborne Uniforms, Insignia & Equipment in World War II

A last minute struggle to fit his arm through the other L-strap. Note the shortened tanker holster – originally issued as a full length holster, later in the war remaining stocks were shortened, and then issued.

As can be seen with all this kit on it is indeed difficult to simply put one's arm through. The ground sheet carried in the haversack is of a lesser known color as most are khaki – other colors being black, and tan.

Left: Finally his arm is through, though not with out some effort. Note the jump wing on the shoulder of the Denison smock. Paratroops wore them on the right sleeve 6" from the shoulder seam. Center: Of interest is the way in which the Brace Attachment is attached to the waist belt – this can be seen just above the Officer's Satchel. This Brace Attachment allows the braces (shoulder straps) to be worn without the use of basic pouches. Also note the revolver cleaning rod in the holster. This was put through the barrel with a cloth patch and then run through the end. The Officer is reaching for the hook of the L-strap to secure it in its keeper which is the top slit of the Brace Attachment. Right: Note the brass buckle sewn to the side of the Haversack. This allows the Haversack to be worn on the waist which would be secured on each side by the shoulder braces. This was done by members of the Air Landing Brigade in Arnhem – as they are Glider troops they would not wear their haversacks on their backs as they must sit flush to the inside of the glider. This was also done by line infantry as they would wear their Large Pack upon their backs (which required the use of the L-straps), thus the haversack was worn on the side, at the waist. Note Col. Sweeney (Oxf. & Bucks) remembers wearing his haversack in both positions – on the back, and at the side.

Chapter One: The Parachute Brigade

A perfect late war look with early war helmet net (1st Airborne), third pattern helmet, face veil, and toggle rope. Note that the pull tie is missing from the Denison's zipper – this tie enabled it to be opened and closed with ease.

The Pathfinders

Each Division had eighteen Parachute Pathfinder teams. Each team was made up of one or two Officers and nine to fourteen other ranks. Duties of the Pathfinders were to mark the DZ/LZ with colored cloth panels, smoke canisters and Eureka Beacons. A Eureka beacon was a ground-to-air radio beacon that sent homing signals to a Rebbeca receiver fitted to the transport aircraft. Colored cloth panels were lain out in the shape of a letter 'T' and also the letter 'X'. These panels would also have special battery powered holophane lamps. A green lamp was placed at the base (bottom) of the letter and an orange lamp was placed at the top of each letter. A member of each stick (section) would flash a Drop Zone code letter using the lamps. This coded letter would be for specific transport aircraft, as each group had their own particular destination and objective. These recognition devices aided the troop and supply transports in dropping their cargo in their proper locations. These men were the 'first troops in' occupied territory. Each team was reinforced by a group of 'protective' personnel. These 'protectors' are to set up a defensive perimeter around the Pathfinders, allowing them to go about their duties (hopefully unhampered by having to engage the enemy while signaling the approaching troop carriers).

There were three pathfinder units, these were the 21st Independent Parachute Company, 22nd Independent Parachute Company, and 23rd Independent Parachute Platoon. About sixty percent of both the 21st and 22nd IPC members were German speaking Jews who had escaped the Nazis. The 21st IPC were attached to the 1st Airborne Division for both Sicily and Arnhem. The 22nd IPC were attached to the 6th Airborne Division for both Normandy and the Rhine Crossing. The 23rd IPP were attached to the 2nd Independent Parachute Brigade for Italy, Southern France, and Greece.

Right: This back view shows how the toggle rope was secured. Shown is the proper Airborne toggle rope – there were thinner variants that were more commonly issued to Commando and Glider Pilots. This soldier is a good reason the LONG web braces were introduced as he is tall and NORMALS would not fit. The early entrenching tool handle is carried here as was done throughout the war – more so by the 1st Division. There is also a good view of the MKIII Light Weight Respirator. Also of interest is the piece of string tied from the canteen to the canteen cork. It was secured via a metal hooked cap that was attached to the top of the cork – corks were replaced as they wore out. The Denison here is a superb 2nd pattern example. Lastly, note how the web chin strap comes down in an inverted V around the back of the neck, and then down and around the ears to be secured at the jaw.

For King and Country: British Airborne Uniforms, Insignia & Equipment in World War II

Left: A very worn 1st pattern Denison. As they were hand painted with ryte dyes, the dyes would eventually fade with wear. Of interest is that most first patterns were converted to full zip after the war. It is rare to see one that has not been so converted – this smock has not been converted and is a rare find. Note also the thinner toggle rope, and a Tankers holster which was very popular among the Airborne, especially 1st Division troops. Attached to the revolver is its lanyard. This Para is armed with the MK V Sten which first saw wide use in Normandy – from this point and on during the war, it was rare to see an earlier model Sten in use by the Airborne. Just in view is the bellows map pocket of the Parachutists trousers. In the background is an Airborne jeep loaded with medical supplies, an Airborne folding stretcher, and a PIAT. One last note is the Mickey Mouse camouflage scheme painted on the Airborne trailer. The British manufactured their own jeep trailers which were box shaped, and were able to carry more equipment than U.S. made trailers. Right: Back view of the trooper showing the early E tool handle, and the Revolver lanyard dangling to the right. Note how nicely the webbing is blancoed and how the brass buckles are polished – it was regulation to blanco the webbing for camouflage. The ground sheet is secured and carried inside of the haversack. Note also that the smock's 'Ape tail' is not secured to the lower back of the smock – this is what makes a 1st pattern smock different from a 2nd pattern. Upon landing the Para would usually undo the tail and allow it to hang loose, thus the nick name 'Ape tail'. Just at the right elbow is a pistol ammo pouch attached to the waist belt.

Left: Rear view of the experimental body armour, which was only issued to the Recce Squadron Drivers, some Glider Pilots and members of the Polish Brigade on a trial basis for the Arnhem operation. Note the stainless steel snaps on the ape tail as worn by the officer in body armor – this is a tell tale sign of a post-war smock, as all war time smocks have solid brass snaps front and back. The male snaps of the post-war smocks are as seen here in steel (the female head snap of this smock is brass). The large web bag/pouch hanging off the waist is the officer's satchel. This is an issue piece of kit for an officer – in it he would carry important papers and such. This is a good shot of the MK V Sten, considered to be the best made and most dependable version of the Sten. You can just see the officers rank slip ons, which are fitted over the Denisons shoulder straps, and feature plastic economy pips as opposed to worsted (woven) pips.

Chapter II

The Airlanding Brigade

Less known, and lesser appreciated were the Airlanding Brigade. Arriving into battle by glider the airlanding unit was larger in number and heavier armed and equipped than a parachute brigade. Their role was to support and reinforce the Parachute Brigades. There were two Parachute Brigades, and one Airlanding Brigade in each Airborne Division. An Airlanding Brigade consisted of a HQ and three infantry battalions. Attached to each Brigade was an Engineer Field Company and an Airlanding Field Ambulance. The first Airlanding unit was formed in October of 1941, the 31st Independent Brigade Group which was later redesignated the 1st Air Landing Brigade. The 1st Airlanding Brigade became part of the 1st Airborne Division, and in May, 1943 the 6th Airlanding Brigade was formed and added to the newer 6th Airborne Division.

Those Who Served
Pvt. Robert Stokes, 2nd Bn. Ox & Bucks Light Infantry, 6th Airborne

From letter dated 4/98:

Enlisted 12th February 1942, my 19th birthday, traveled to Bodmin barracks and stayed for five months training. Transferred to Somerset Light Infantry 3rd July 1942 where I volunteered for service with the newly formed 6th Airborne forces on 10th October 1942. Throughout the early months of 1944 we were subjected to intensive training in airborne warfare, this was in preparation for the invasion of Normandy. On 4th April we took part in an exercise where the whole division was taken into the air by the RAF, unknown to us, this was a dress rehearsal for D-Day 6th of June 1944.

On 6th June 1944 we assembled on the runway of Harwell Aerodrome. We seemed to stagger into the air with a heavier load than we should have had, for every man had taken a few extra grenades and rounds of ammunition than had been allocated (issued). We crossed the coast near Worthing and headed towards Normandy. The channel below was packed with boats of all sizes. We approached our landing zone and could see that Major Howard's party had landed dead on target (Major Howard had taken off at 2256 hours and by 0016 hours had landed atop the bridge. I left on the afternoon of D-Day and landed on the left flank of the bridge [Pegasus bridge])!

We cast off (released the tow line from the glider to the towing aircraft) about a mile from our target in the area of Ranville. On our release it became deadly quiet ... and the sudden dipping of the glider's nose made our ears pop and we shouted as loud as we could to release the pressure in our ears. Hopefully any enemy troops within earshot would have heard this and been frightened out of their wits!

The trees and hedgerows seemed to be coming at us at an alarming speed! The field in which we landed was an area we had studied in some detail (sand tables had been set up to aid the troops with their tasks upon landing), unfortunately we had expected a hard surface to land on, but the farmer had decided to plough it just the day before! This caused the gliders to land nose first and crash head on at 70 knots and fully loaded with cargo, they became a pile of matchwood on contact.

The next few hours are very vague, working to unload the gliders under German machine gun fire. Tragically our glider pilot was killed on landing, and sadly we were forced to leave him behind and get to our target (objective) to get dug in, ready for the German counterattack which we all knew would certainly come. This was surely our longest day.

We held our position until 31st August, when we advanced to the River Seine. By September we were back at Bulford Camp (UK). Christmas 1944 we were rushed to the Ardennes to stem the German offensive. We later returned to Bulford on 28th February 1945. We started training for Operation Varsity, the crossing of the Rhine, where we unfortunately lost half the regiment on landing (German anti-aircraft was very active that day). We reached the Wismar, on the Baltic Coast, when the war ended on 8th May 1945, by 20th May we were back in Bulford.

For King and Country: British Airborne Uniforms, Insignia & Equipment in World War II

The Glider Infantry

An Airlanding Battalion was not made up from volunteers as with the paratroops, but rather was redesignated Airlanding by, for example, being put on parade (formation) and then told that the unit was becoming an Airlanding Battalion. Anyone who wanted to transfer could then do so. Pay for the Glider troops was one extra shilling for other ranks and two extra shillings for Officers. This was 'hazardous duty' pay. Most remained with their mates, who they had gone through training with, not to mention the extra pay which for most went a long way.

The following Line Regiments battalions were redesignated Air Landing:

1st Airborne Division: 1st Battalion Border Regiment, 7th Battalion Kings Own Scottish Borderers, and the 2nd Battalion South Staffords.
6th Airborne Division: 12th Battalion Devons, 2nd Battalion Oxfordshire and Buckinghamshire Light Infantry, and the 1st Battalion Royal Ulster Rifles.

The Airlanding Battalion was larger than a Parachute Battalion being made up of 800 of all ranks (officers and enlisted). Each Battalion was comprised of a Battalion Head Quarters, a Head Quarters Company which consisted of a Reconnaissance platoon, a Pioneer platoon, four Rifle companies and a Support company ('S' Company). An 'S' Company consisted of a machine gun platoon with four Vickers machine guns, a mortar platoon equipped with four 3" mortars, and an anti-tank platoon with eight 6 pounder anti-tank guns. Each Rifle company consisted of 6 officers and 150 other ranks. This was broken down into a company HQ, and 4 platoons of 36 men each, and a 3" mortar section. There was also a Carrier platoon (Bren carrier).

Members of the Air Landing Brigade enjoy a break. Note how the shoulder titles are censored for security reasons, as was common practice. Below the blotted out title are the printed economy Pegasus flash and Airborne strip. Of interest is that all three men wear the 1942 fiber rim helmet, which was discontinued and replaced by the all steel shell in 1943, and also featured the same leather chin strap. The inverted chevron on the sleeve of the Para on the right is a Good Conduct award. These were worn by other ranks up to the rank of LCpl. Ranks Cpl. and higher did not wear them as to hold such rank meant good conduct was expected. The early pattern of battledress is worn by all three, this being either the Battledress Serge or the Battledress Serge 1940 Pattern. (The differences are covered in detail in Chapter 5). They also wear waist belts by regulation, as seen by the man in the middle. Notice how well polished the brass buckle is.

A South Staffs man on the LZ. He wears the issue maroon beret with his regimental cap badge, as his battalion was redesignated Air Landing. Had he been a member of the Parachute Regiment, he would not be able to wear his regimental badge and would, by regulation, wear the Parachute Regiment badge. His toggle rope is the typical thick issue variant. The white tie string on his shoulder strap is a shell dressing – this was common practice for the Arnhem battle.

Right: Pictured here is Lieutenant Colonel Robert Payton Reid of the King's Own Scottish Borderers. Of interest is the embroidered Airborne strip. Above the Pegasus flash is the clan Tartan, Leslie Tartan of the K.O.S.B. The 7th Battalion K.O.S.B. wore these tartan flashes in place of shoulder titles. The Austerity pattern Battledress is worn, as there are no pleated breast pockets, and the buttons are exposed. Also note the regimental cap badge on his maroon beret, and as an officer he is, by regulation, to wear a tie. Other ranks wear not so authorized, and in fact did not 'officially' wear ties until late 1944.

Chapter Two: The Airlanding Brigade

Above: The Border regiment on Parade – all members wear the regimental cap badge on their berets, and the glider badge on their right sleeves. These are known as Qualification Badges, and are authorized to be worn upon qualifying for that particular badge. An example is the parachute 'light bulb' Jump Qualification badge seen above the glider, second on the left. A clasp knife lanyard can be seen on the young trooper, second on the right. Waist belts are blancoed to regulation. The strips of material on the shoulder straps are colored per company, and each company had their own colored shoulder slip on. These were introduced after the Battalion became part of the Air Landing Brigade (note that it was one battalion of the regiment that went Air Landing and not the entire regiment, as was the same with all Air Landing Battalions). The Border Company slip on colors were: HQ company yellow, A company green, B company white (then green to avoid confusion with officer cadets), C company red, D company pale blue, Support company dark blue, and T (later R) company black. Below: A Bren Gun team secures the surrounding woods of the LZ. Of interest is the shell dressing hanging from the shoulder of the Para (as he's wearing Parachutist trousers – they were not issued to glider troops). The Rifleman is from the 7th Battalion K.O.S.B. – note the tartan flash backing behind his cap badge. This was not done, as the badge was worn upon their berets without a tartan backing.

For King and Country: British Airborne Uniforms, Insignia & Equipment in World War II

Above: S. Staffs Sten Gunner making his way off the LZ and into surrounding cover. Of interest is the Bergen Rucksack underneath his right arm and the Bren Auxiliary pouches which are attached to the Bergen (one pouch on each side). Bergens were issued to Air Landing bicycle troops. On the front of their bikes was a special bicycle rack, and was used to secure the Bergen while mounted. The object held in place on the Sten Bandoleer is a British made box Sten magazine loader. The gunner is armed with a MK II Sten. Note that the rank chevron is worn but only on the right sleeve – this was common practice among the Airborne, possibly to conceal rank from enemy snipers. Note the 2nd pattern smock (as it features button tab wrist closures).

Above right: S.Staffs Sten Gunner loads his weapon. Note that he has stuffed the excess cuff material back up into the sleeves of his smock – a personal preference.

Right: The string around the right shoulder is a clasp knife lanyard. Note the fresh color of this smock. This again drives home the point of how each Denison is of a character of its own. Some sort of elastic band seems to have been sewn into the cuff on the very bottom, and was done by the individual and not as issued.

The Early days – a classic shot from a training exercise. This is an extreme close up of a Hotspur glider. The Glider Pilot wears the early leather flying helmet. The Sten Gunner is armed with a MKII Sten and wears the 1942 Fibre Rim helmet. Rank chevrons are seen on his battle dress and were worn on both sleeves of the BD blouse. His shoulder title is censored for security reasons. The Hotspur was the very first glider to be available for training purposes. It carried eight men – one Pilot and seven troops.

32

Chapter Two: The Airlanding Brigade

Above: A Vickers crew minus one man (this gun was meant to be manned by between three and five men), as this is a heavy and cumbersome weapon to move – not to mention the tripod, water can, spare and repair kit, cleaning kit, and ammunition. Therefore this gun could not be moved on the double quick. Please note that the water condenser can and condenser hose are missing from this group of photos. To the left you see the .303 ammunition box, and further back is the spare part and repair kit. This crew has just set up their weapon to cover the LZ and protect it from counterattacking Germans. They now check to see that the weapon is in working order, as some weapons were damaged upon landing. Note the canvas jacket (barrel) cover was to enable it to be carried, otherwise one could burn himself from the recently fired weapon. Right: Here the belt has been fed through the gun and is ready to be fired. Note both troopers are in full order (wearing their full allotment of kit).

The Vickers Team, Support Company, Border Regiment, 1st Airborne Division

The Vickers was widely used by the Airborne troops. Though heavy and cumbersome, it was the only heavy machine gun available to His Majesty's services. The Vickers saw service with the Airborne throughout the war. There was also the Vickers K gun which had a higher rate of fire, and for the Airborne it remained a mounted weapon. The Recce Squadron made good use of them and mounted them on their jeeps. Commandos, to my knowledge were the only troops to actually carry the K gun into battle by hand.

Left: Here the leather Vickers cleaning kit is worn over the assistant gunner's shoulder. Note the No.7 bayonet which was introduced in early 1945 in time for the Rhine Crossing (and actually should not be in this shot). His E Tool handle is the late war variant as it has the metal point on the end – this was for mounting the spike bayonet and was to be used as a mine probe. Also of interest are the hob nails on his boots, which makes these the late war variant, 1944-1945. As the war went on the nail studs decreased in size. At the beginning of the war they were 1/2" in circumference. Each pair of boots had a heel cleat (seen here), and a toe plate for better traction. The Gunner wears a compass pouch on his waist belt between his arm and haversack – as he is equipped with a smaller variant toggle rope, he has secured it in this fashion. Note also that the gunner is wearing jump wings. It is possible for a soldier to be transferred from Para to glider, and from glider to Para, yet this is an exception rather than common practice. Below: At this point the crew has decided to clean the gun as dirt and dust have got into the gun from the landing. If the weapon is dirty it could jam. The incoming gliders will be depending on those on the ground to keep the landing areas safe and secure.

Left: The firing mechanism is taken out and cleaned thoroughly. Note the labeling on the wooden repair kit below.

For King and Country: British Airborne Uniforms, Insignia & Equipment in World War II

Left: The crew oils the gun. The small tin to the left is an oil can. The lid unscrews and has a oil brush attached to the other end of the it. The Gunner has unscrewed the left handle top, which is also an oiler. Right: A clearer view of the inside of the gun. Of interest is the spooling nub of the crewman's beret. By placing a finger on the very top of the beret a small ball in the fabric can be felt. This is the spooling nub, and the tell tale sign of a war time beret rather than and a post-war piece. Berets made now are not spun like they were during the war. The berets were spun by knitting factories on large needles. Starting from the top and then spun down to the crown, where the leather sweat band was later attached.

Above: The gun is clean and ready for action. The wooden lid on the ground just below the assistant gunner is the lid of the .303 ammunition box now in use.

Above right: It is decided to move the gun to a better position as few guns are working and or serviceable due to their being damaged upon landing. This gun will have to do the work of others and fill in the gaps. Note the beret of the Gunner – war time berets are larger and will cover one's ear. This beret has either been shrunk from post-war use (this was common practice to look sharp. A Para would wear his beret in the shower, soak it, and then the beret would shrink and conform to the his head. This was not a war time practice), or it is in fact a post-war beret. It's important that the collector know this. War time berets looked like large soup plates upon the heads of the Paras. Once worn, it could then be pulled down to the ear lobe. This is standard with any war time British beret. In the post-war years they were shortened by about 1".

Right: Details of the Vickers. This was the heavy machine gun of the Airborne. There was a Vickers 'K' gun which was mounted upon jeeps (this was done by for example by the Recce Squadron). There was also a hand carried 'K' gun used by the Commandos. The Airborne did not use this variant.

34

Chapter Two: The Airlanding Brigade

Left: Of note is how all the webbing is camouflaged with blanco. Note the handle of the bayonet. It is interesting that England would make such an intricate weapon when they had the simple spike bayonet already in service. Again, the black leather pouch carried by the assistant gunner is the Vickers cleaning kit. Right: Note the detail of the tripod's clutch plate handles which adjust the tripod legs, and the elevating wheel which determines the elevation at which the gun will fire at. This tripod weighs 40 pounds.

Note the chamois lined chin cup. This was done for added comfort as the raw leather would rub a man's chin raw. Also of interest is the haversack and L-strap attachment. Note how the haversack is buckled to the shoulder strap (L-strap). The ground sheet is rolled up and fits into the top of the haversack.

A close up of an Air Landing Rifleman equipped with the large toggle rope, camouflaged face veil, 2nd pattern Denison, and 3rd pattern, all steel jump helmet with leather chin strap. – note this chin strap is riveted. Some are both riveted and sewn., others simply sewn. The helmet net is the late war 1944-45 type of single color. There are many colors, such as brown, dark olive, olive as shown, dark chocolate, and split camouflage (forest green and chocolate).

35

Chapter III

Corps and Administrative Departments

In the planning stages of the airborne and through lessons learned in the field both in training and in combat, it was deemed vital to supplement an airborne division with their won support troops and services like that of an infantry division yet specially tailored to the needs and design of an airborne force. These men arrived into battle via both parachute and glider and were attached to each of the Brigades within the Divisions. They are the Corps and Administrative troops.

R.A.Ch.D. (Royal Army Chaplains Department)
The Chaplains provided support and at times, when necessity demanded, assisted the RAMC in caring for the wounded and dying. Many a brave padre jumped into battle to do his part. I'm sure many of the troops found their service of great comfort.

C.M.P. (Corps of Military Police)
Military policemen directed troops to their 'collection' points, provided field security for Divisional HQ, took charge of delivered prisoners and gave them safe custody under the rules of the Geneva Convention, provided safety for Civilians caught in the 'fray' of battle, and supplied guidance to routes from the DZ/LZ to the objectives. And of course, they directed traffic.

R.A.C. (Royal Armoured Corps)
The Armoured Corps provided special Reconnaissance groups who employed the use of Tetrarch light tanks, Bren Carriers, Dingo Scout Cars, and Locust Light tanks (Rhine Crossing only). The R.A.C. did not come into use until the Normandy Invasion of June 6, 1944. There all the vehicles were taken into the field. The R.A.C. was not attached to the 1st Division. (The 1st Airborne Recce Squadron was not part of the R.A.C.) The Bren carriers that were used at Arnhem were attached to the various Airlanding Battalions. The R.A.C. attached to 6th Airborne took part in the Rhine crossing – those that survived the enormous flak that is. "The Rhine Crossing was an embarrassment. We were shot out of the sky!," as quoted by Tish Rayner a Rifleman of the Oxf & Bucks, 6th Airborne Division.

R.A. (Royal Artillery)
The R.A. provided artillery support for the division. This was comprised of the following: the 75mm howitzer (U.S. made) and the 25 pounder field gun. Anti-tank artillery being the 6 Pounder and 17 Pounders. Also anti-aircraft guns, the 20MM Oerlikon and 40MM Bofors. The R.A. also provided Forward Observation Officers with the line infantry's batteries. The FOOs were able to call in added support for the Airborne from the line units. Line units, regiments and divisions had heavier

A member of the R.A.S.C. lights up and enjoys a smoke on the LZ. Note the brass RASC cap badge on his beret – both plastic and brass were worn. His smock is a first pattern, distinguished by the tapered sleeve and knit cuff. As can be seen there is a major difference between this issue cuff and a hose top that has been added after issued. Also note the manner in which the camouflage has faded. First patterns were hand painted with special color dyes, as the 2nd patterns were screen printed.

Chapter Three: Corps and Administrative Departments

guns at their disposal as the airborne was limited to what it could transport via glider.

R.E. (Royal Engineers)
The Engineers provided demolition work and logistical support. Also rcfcrrcd to as Sappers, these men were employed to destroy enemy defensive positions, clear minefields and booby traps, and build and repair roadways and bridges. Two Parachute Engineer Squadrons were attached to a Parachute Brigade. A Parachute Squadron consisted of a HQ and three troops of thirty men each. Each troop was commanded by an officer, usually the rank of Captain. A troop had three sections consisting of a subaltern (1st or 2nd lieutenant) and nine other ranks.

A field Company of Engineers was attached to each Airlanding Brigade and was comprised of a company HQ and three platoons. Each platoon had one captain, a subaltern as the captains adjutant, a platoon Sergeant, and three eight man sections each led by a Corporal.

R.S. (Royal Signals)
Signalmen were responsible for maintaining and operating radio and line communication between Divisional HQ, Brigade, Battalion and Company. They were also responsible for calling in artillery and air support from outside of the Division. Each brigade had its own Signals sections made up of two officers and seventy-three other ranks. Sections were broken up into smaller groups of radio operators, linemen, and cipher operators.

Here is a group of prisoners taken at Arnhem/Oosterbeek. The soldier in the foreground is a Signaler with his special Signals pouch at his side. He wears the 3rd pattern helmet. To the left is a wounded Para whose smock has been cut by Aidmen (medics) to better dress his wounds. Note he is not wearing a collared shirt but rather the Denison collar and what remains of the smock on that side of his body. The sleeve has also been cut away.

Those Who Served
Harry Howlett, Royal Corps of Signals

I joined June 1st, 1940 aged 18 years old. Dunkirk was on and things looked bad. I left a note on the table which said 'Mum I have gone to join the army, see you soon!!' I volunteered for the Kings Royal Rifles as my Brother was also in this unit. We mainly guarded aerodromes for some 18 months until the RAF Regiment took over this job.

I was next posted to the Royal Signals. We were sent to invade North Africa. We were the advance communications so we were in the front of it. Sometimes we went too far and had to do a quick about face! It was here in Africa that volunteers were wanted for the Parachute Regiment so myself and others volunteered. We had just missed Sicily and were sent to Italy. We liberated com wire from the Italians and Germans as we had run short. We pushed up to Foggia and were then relieved. We were sent back to England to prepare for Normandy but this was given to 6th Airborne. So you know what happened next! We spent our time in Holland dodging mortars and repairing phone cable! I was fortunate and managed to get back over the river safely.

(Author - Harry does not have any photos of himself from his army days. As with many Veterans, such little treasures are misplaced or thrown out by a careless family member. Little bits of history that are destroyed or cast aside like the daily trash. As a note, Most of the 2,095 men who escaped from Arnhem crossed the river by boat. Many who tried to swim, drowned. If you take this into consideration that after fighting for nine days, without food, and or without sleep, and the fact that then the river was very fast flowing, it was very difficult to cross. Also of interest is the fact that most Englishmen at the time were not accustomed to swimming as we are now in this day and age.)

For King and Country: British Airborne Uniforms, Insignia & Equipment in World War II

Here a Signals Section ensures that the area is clear of the enemy before setting up a temporary signals station. The Signalman on the left wears a 2nd pattern smock, as noted by the wrist tabs. Also note how loose his sleeves are, this is one of the big difference between patterns. The Signalman on the right wears a 2nd pattern 1946 smock with knit cuffs. Both are part of the Parachute Signals Section. Both have the P-37 Sten Pouches – these pouches are slotted inside to hold the magazines, three per pouch. Many of these Sten Pouches are converted from shortened Lanchester pouches. Lanchester magazines hold 50 rounds, therefore the pouches were originally longer. Both carry the No.38 Radio Set, and are armed with MK V Stens which was the weapon commonly issued to specialists. In the background is an abandoned Bren gun – most likely as its gunner ran out of ammunition.

A Signaler listens to a group of Germans who have tapped into their lines. He wears the Signals badge upon his beret and is in 'shirt sleeve' order (he's not wearing his BD blouse, simply just his collarless shirt).

A good look at the second Signaler as he checks his frequencies – still trying to reach Divisional HQ. Around his neck is the throat microphone. This was used in place of the bakalite No.2 Microphone. For troops on the move the throat microphone was ideal.

Right: Here is the manpack (backpack carrier) for the No.38 set. The battery fits into the left slotted compartment, and the radio on the right. The two snaps to the left of the carrier is so the battery can be changed without removing it from its pocket. The round impression on the bottom end of the MKIII respirator is the anti gas cleaning tin which contains small bits of cloth which were used to clean gas residue from the eye lenses of the gas mask.

Chapter Three: Corps and Administrative Departments

Note how the web breast carrier is attached to the brass hook of the L-strap. To the right is a loop and buckle. This is a standard signals shoulder strap which was used with assorted signals equipment. To the lower right is the signals satchel in which is carried the head set, microphone, and battery.

Shown here is a Signaler and guard. The guard keeps an eye out for the enemy and frees the signaler to care to his duties. Of interest, in the signaler's right hand is a switcher which controls the volume and is an on/off switch – talk on, listen off, much like a present day hand held walkie-talkie. Note the leather anklet straps on the Para providing security. These rotted when wet and exposed to the elements. The all web anklet proved to be more durable.

Note the vertical indentations of the P-37 web pouches. These are the slots for the Sten magazines. Also notice the clean cut edge of the signaler's 2nd pattern smock, as issued and unmodified with wool knit cuffs. Around his neck is the throat microphone.

Above: Note the hand switcher and volume control. The small tins on top of the wooden crate are boiled sweets (issued hard candy). To the left of the No.38 set is a metal plate with instruction sheet which was printed on the metal sheet for durability. This was carried inside the web radio bag. To the right, also on the crate is a second head set and clip on microphone.
Right: Note the switcher atop the haversack in the middle, against the crate. In front of this is the signals satchel. Atop of the satchel is the volume control for the No.38 set in the web breast carrier. The Signaler on the right adjusts his throat microphone.

For King and Country: British Airborne Uniforms, Insignia & Equipment in World War II

The blue cord is for the throat microphone which is plugged into the volume control box. This signaler is adjusting his throat microphone for a secure fit.

Another look.

Battalion level switchboard complete with field phones and typewriter for sending and receiving messages. The large gray unit is a No.19 radio set, and atop that is a No.38 set.

From front to back: a .303 ammunition box, two spare valve boxes (foam lined for protection), radio tubes in the wooden boxes with leather carry straps and lastly a signals lamp-inside the metal box with web lid. To the left hanging off the bed is the Airborne Sleeping bag.

Chapter Three: Corps and Administrative Departments

Above: More of the Regimental/Battalion Signals post: spare batteries to the right in the card board box, field manuals, a Pattern 37 shell dressing bag filled with dressings, map cases and signals satchels. Right: To the left a No.38 radio set and web Manpack, a tin of rations, and a MK V Sten held in place by a web weapons rack which is secured by fitting over a tent post. Behind is a large issue tent. Tents were not part of the Paras kit as they were lightly equipped for mobility. As for the HQ section, tarps and tents were brought in by glider. The front of the tent is closed by the twine loops and wooden toggles that are sewn to the other tent half.

Right: An Airborne Pigeon Carrier – pigeons were used in both Normandy and Holland by the RS to deliver messages in the field (in combat). The Para wears what looks like a 1942 fiber rim helmet as one can just make out the rim underneath the netting. Note the short issue knit cuff of his smock. Of interest is the MK II Sten carried in his harness. Note the plate which can be set and locked-in so the weapon can be fired without the metal stock. This extra plate would be carried in the Sten bayonet frog – the frog had a special pocket to hold the plate. Also of interest are his jump wings on his sleeve. These have been closely cropped as was common practice. Behind him is his static line which once inside the plane he will hold onto until given the command, by his jump master, "Stand Up, Hook Up." This will take place while over the DZ. Above: A Type F Signals Drop Container. Packed inside is a No.22 set and two signals satchels which contained the headsets, microphones, and batteries. This container is specially padded and reinforced to sustain a hard landing as this entire unit would be dropped by parachute.

41

R.A.M.C. (Royal Army Medical Corps)

The R.A.M.C. provided aid, set up and ran the various casualty clearing stations and operating theaters. Each battalion had its own medical staff and medical platoons were attached to each battalion. Troops were selected by members of the R.A.M.C. to serve as stretcher bearers – they would wear the same brassard (arm band) as the doctors and aidmen.

The Field Ambulance was an early version of the modern MASH unit, and provided care for more serious medical situations. It was supplied and stocked with specially designed medical and surgical equipment, and blood plasma, all of which was delivered by both parachute and by glider. Each brigade had its own FA which came in by glider and set up operating theaters and the various CS (casualty stations), CCS (Casualty Clearing Station), and MDS (Main Dressing Station).

A Parachute Field Ambulance provided medical support for the Parachute Brigade. Each had about nine officers and one hundred other ranks broken down into a HQ element, two surgical teams, and four clearing sections. An Airlanding F.A. was larger in number than the P.F.A. and better equipped as they arrived by glider and were able to bring in more equipment and supplies. An Airlanding F.A. would be equipped with jeeps and occasionally a Bren carrier. These men were in the thick of it along with everyone else.

An aidman (medic) operates on a wounded Para. Of interest are his surgeons implements roll on the wounded man's leg, and the wicker Field Ambulance Pannier. Also note the surgery towel, and 1st pattern folding airborne stretcher (balled handles, as the 2nd pattern were without the balled ends). Line Infantry stretchers were non-folding. The idea was to keep them compact for transport in gliders.

Shown here are the partial contents of the inside upper tray of the Field Ambulance Pannier. From left to right: shell dressings, first field dressings, ointments, syringes, gauze, sterile bowls, where alcohol will be poured in and instruments are kept clean while operating, foot powder, and yet more shell and first field dressings. To the far right is the end of the folding airborne stretcher.

To the left and in front of the chest is the Shell Dressing bag. Of interest are the instruments on the surgery towel. The Aidman is finishing up as a Para across from him lends a hand by preparing to apply a first field dressing.

Left: Pressure is applied and the dressing is set in place.

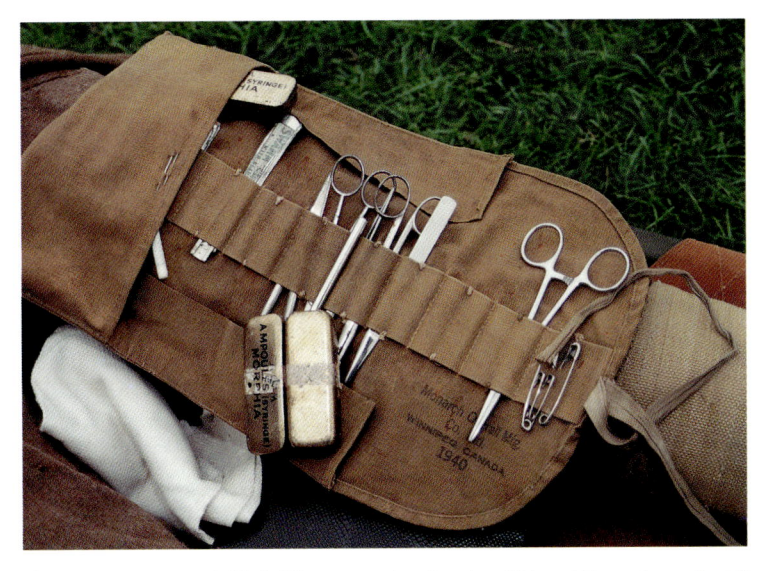

Chapter Three: Corps and Administrative Departments

The surgeon's instrument hold all. Of interest are the safety pins which could be used as make shift stitches. Needles are seen stuck through the top of the hold-all pocket.

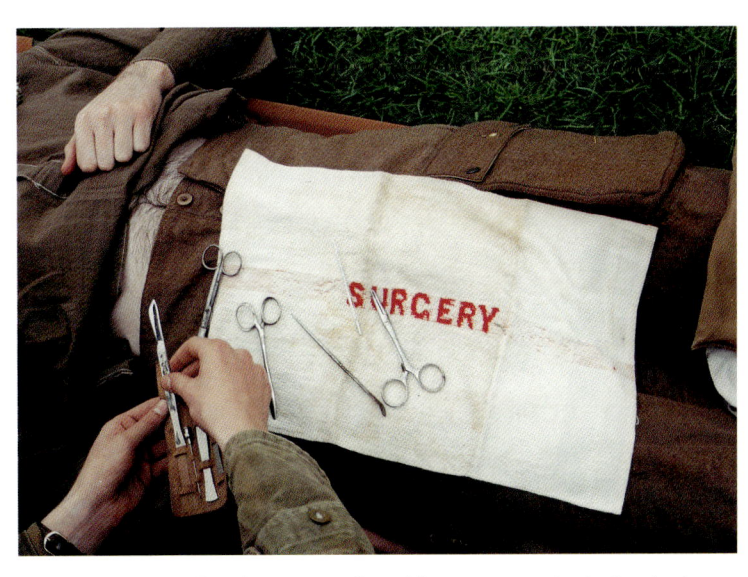

This surgery towel was to keep the instruments clean while in use as opposed to simply setting them on the man's trousers which would be soiled and therefore unsanitary.

Right: Underneath the shell dressing bag is a metal first aid box dated and broad arrow marked – meaning that all pieces of equipment made for the services (Army, Navy, etc.) are proof stamped. This stamp is the sign that this piece of equipment has been inspected and approved of by the War Department. Also of interest is the 1st pattern smock worn by the Para to the right. It is a 1st pattern as its colors seem to have faded, and it also lacks the Ape Tail snaps on the lower back.

A Company Aid Post. From left to right: Large pack containing medical supplies, Bren spare parts kit (out of place, possibly left by a wounded Para, a Bergen Rucksack, atop the rucksack is an Officers SD (Service Dress) cap with Officers RAMC cap badge, behind this is an airborne folding stretcher. On the camp bed (cot) are the officer's personal kit: haversack, shirt, Hold All, rations and mess tin. In the middle is a Shell Dressing bag which held 6 dressings, another large pack filled with medical supplies, and a 2nd pattern Denison with added and non issue hose tops.

43

Reconnaissance

Know as the 'Recce', their job was to scout ahead of the division, ascertaining and providing information of enemy strengths, positions and intent. The unit traveled in special jeeps (transported into battle via glider), usually equipped with wireless sets. Some of the troopers would travel with their jeeps in the glider while other troopers arrived by parachute. The jeeps were not armored, so their occupants relied solely upon their firepower and mobility to get them out of trouble. Each jeep featured a mounted Vickers machine gun known as the Vickers 'K' gun. This was a drum fed aircraft weapon with a very high rate of fire thus many a spare ammunition drum had to be carried. These jeeps were specially modified for glider transport and were unloaded on the drop-zone by means of removing the glider's tail section and driving it down a pair of special ramps. Many gliders were damaged during assaults either by enemy fire and/or a harsh landing, thus making it difficult, if not impossible, to remove the vehicle.

As per Churchill's request for the formation of a corps of parachutists, the 1st Airborne Reconnaissance Squadron was formed in 1941 as a part of the new British 1st Airborne Division. Originally known as the "1st Airlanding Reconnaissance Squadron" they served in North Africa and took part in the invasion of Italy in 1943. After its return to England in early 1944, it became fully parachute trained and was renamed the "1st Airborne Reconnaissance Squadron." Training for other operations was now undertaken, though the squadron did not take part in the Normandy invasion of June 6, 1944. Following a long succession of canceled operations (some were actually called off as the men boarded their aircraft), the squadron took part in Operation Market Garden, on September 17, 1944. There they fought for their lives, alongside the rest of the Division, as not all went to plan. Those not taken prisoner escaped by swimming the Rhine and were later sent to special rest camps to 'rest and refit' so they would fight yet another day (however Arnhem was the last combat they saw during the war).

The Recce were then moved to Norway at the end of the war where they took part in an operation to disarm German troops and to assist the resistance movement in the identification and segregation of known SS personnel and war criminals. The Squadron served in Palestine before being disbanded in 1946.

Shoulder insignia of the Recce, embroidered shoulder title, jump wing, and printed economy Pegasus and Airborne strip. The rank chevron is that of a Lance Corporal. This being the equivalent of a U.S. Corporal. A LCpl. was a junior section leader (assistant squad leader).

Here Recce troops show their spirit prior to leaving for Arnhem. Of interest is the plastic economy Recce cap badge worn by the middle trooper in shirt sleeve order. Also note the Vickers firing ring – this is the large metal rod with the ring. This was placed in front of the weapon to keep it on its fixed line of fire.

Chapter Three: Corps and Administrative Departments

Shown here is the Recce Jeep – of interest are the Vickers 'K' gun and its mount. Note that the windscreen has been removed, and a special frame with webbing straps has been added so that needed .303 Bren chests may be stored there. The Trailer, 10 cwt, Lightweight, general service, was employed by all units of the British Army. Because of its high sides it is said that it could float. At the rear was a tow hook which enabled another trailer to be towed behind.

Above: Here a trooper checks and double checks prior to the big show! Everything must be in working order least it fail when it is needed most. Note how 'stripped' the jeep looks, this was done to lighten the load and better fit the jeep into the glider.

Right: Note the many 'Airborne alterations': the Bonnet ammunition fittings: fittings on the front bonnet for the carriage of ammunition were added, and were of a two piece construction frame as was used by the Recce Squadron. The Detachable Steering Wheel: To get a jeep into a Horsa glider the steering wheel was made detachable by a wingnut. The horn button on the wheel was replaced by a standard bakalite pushknob which was mounted on the dash, usually on the left side, next to the blackout light switch. The horn and the panel lights were disconnected during combat. If an ignition key was used it was connected through the panel light switch. Front Bumper: In order to get the jeep into the Horsa the bumper was shortened and reinforced, and a towing hook was added. Jerrican holder: The U.S. made Jerrican holder was removed from the rear and fitted to the floor between the two front seats. Gun clips were added to the right mud guard and on top of the dashboard which would hold a No.4 rifle. Spade and pickaxe brackets were removed to enable the jeep to be loaded easier. The spade and pickaxe were moved and mounted on the front bumper. Vickers "K" gun: A special bracket was added for the gun to be mounted on the upper part of the dash. This was only done by the Recce Squadron. Side and rear handles were removed to ease loading and unloading. Lights: The black-out driving light was removed and both side lights were taken from the grille and placed on the top mud guards. The towing hook was equipped with anti-rattle pads as the security ratchet would at times open up from the twisting of the trailer. A hole was drilled into the hook and ratchet and a pin was placed in the hole, which was attached by a chain. This kept the security ratchet from moving while in motion. And lastly, the Pannier carrier: the rear was removed and a special gate added to enable the cumbersome wicker Panniers to be carried.

For King and Country: British Airborne Uniforms, Insignia & Equipment in World War II

The Pannier gate and tow hook.

Above and below: A trooper has spotted some movement just in the hedge before them and all scramble to get into position. The Bren Gunner prepares to cover his mates as they flank the hedge. The driver has grabbed his rifle from the dash gun clip and scans the hedge for the enemy.

Chapter Three: Corps and Administrative Departments

Right: Recce Sten Gunner on watch. Note the plastic badge – the troops did not like them as they could not be polished. The British Army loved/loves to polish brass. Yet the plastic badge does not stand out or shine in the sunlight, thus not acting as a beacon to enemy troops. He is armed with a MK V Sten and sports the experimental body armour. Note how it flexes with the body. Once the jeeps were lost, he had to fight on, on foot, providing support where he could.

Below: Front full view of the Recce Jeep Driver. Note the foregrip on the MK V Sten – MK IIs and IIIs did not have this feature. Just visible hanging behind his left shoulder is an American walkie-talkie, a BC-611 – this was Lend Lease from America.

Below right: Rear view for the Recce Driver. Of interest is the rear plate of the body armour, and the hob nails of his ammo boots. These are the early war stud nails, being 1/4", and were used from the beginning until about 1943. From then on the studs were smaller and fewer were used per boot – this was done to save materials for the war effort. One last note are the steel rear Ape tail snaps. This is a post-war (1946) smock as WWII smocks feature all solid brass snaps. Otherwise it is very much like a WWII made Denison.

47

For King and Country: British Airborne Uniforms, Insignia & Equipment in World War II

R.A.O.C. (Royal Army Ordnance Corps)
The R.A.O.C. was a cross between the R.A.S.C. and the R.E.M.E., performing similar duties in the field. Their main duty was that of ordnance, this being the issuing of weapons and ammunition, as well as the repair of damaged weapons.

Above left: A drop container full of arms – MK V Stens, loaded Sten magazines, and loaded Bren magazines. Note the still attached parachute webbing harness to the bottom right – still taught, meaning that the chute is still deployed and full of wind. This trooper is in a hurry to get his weapon and magazines! As the R.A.O.C. were mainly concerned with ordnance they would collect all of the ordnance and distribute it to the troops – not to mention getting off the DZ/LZ and out of the line of fire from enemy troops. Above right: Shown here are the drop containers, soon to be loaded on the aircraft. Of interest is how the weapons were wrapped in thick blankets to prevent damage upon landing. Here are a Bren and a No.4 MK I Enfield Rifle which was the fastest bolt action rifle in the world. A trained soldier (British and Commonwealth) could get off all ten rounds in one minute. This is a technique done by squeezing the trigger with the forefinger and operating the bolt with the index finger and thumb. Note the bulb shaped ends of each container. These are called Crash Heads and are designed to absorb the shock of landing, thus helping to prevent less damage to its contents. The container clamps which are the Cradle Arms, officially know as Cradle, Type CLE MK I., can be seen in the middle of the right container. Once this was off the container could be opened. The chute would be attached to the other end as the container descends vertically. The parachute and bag can be seen at the top of the container on the left. There were different colored chutes used as each signified a particular type of ordnance, ammunition, food, and or supplies – colors used were blue, black, yellow, orange, and red.

This is an interesting piece and has not been seen by the author before. As can be seen it holds three M 36 Mills Bombs. There is a small slot on the far right. This looks like a vehicle mounted box.

Metal Bren ammunition chests. Each chest holds 12 magazines. These could be, and were, carried atop a jeep hood, or broken down and distributed among the men to carry in their P-37 basic pouches as they marched by.

A box of .303 ammunition. One type comes in boxes like those pictured. These can be loaded into either Bren magazines, Vickers belts, or into Enfield stripper clips which hold 5 rounds each. Each box is sealed to keep them dry and usable. Another .303 crate, looking much like this opens to reveal loaded bandoleers complete with all rounds on stripper clips. These are then issued to each Rifleman.

Chapter Three: Corps and Administrative Departments

Intelligence Corps
Their role was to provide information about the enemy and prevent the enemy from obtaining information about the division. Interrogation of prisoners provided information that was of use to the division as well. They were directly responsible to Divisional HQ.

R.E.M.E. (Royal Electrical and Mechanical Engineers)
The R.E.M.E. provided maintenance, repairs and spares for the various vehicles, signals equipment, and weapons employed by the airborne.

A.F.P.U. (The Army Film and Photographic Unit)
The Army Film and Photographic Unit took photographs and shot film of the battle. At Arnhem this was Sergeant M. Lewis, G. Walker, and D. Smith.

R.A.S.C. (Royal Army Service Corps)
The R.A.S.C. secured the DZ, kept it free of enemy observation and direct fire, kept the air approaches free of enemy anti-aircraft weapons, kept the approaches correctly marked, offered radio aids to the approach, provided labor and transport for the clearance of the LZ/DZ, and were responsible for the safe collection, storage and distribution of supplies. Of interest is that the highest number of casualties among the troops at Arnhem were those of the R.A.S.C. The R.A.S.C. also provided dispatchers who flew in the transport aircraft and pushed the panniers (wicker supply containers) and drop containers from the aircraft. These troops were known as Air Dispatch.

Special R.A.S.C. companies were formed and deemed L.C.C. (Light Composite Company). The L.C.C. were equipped with jeeps and trailers which arrived by glider. These jeeps and trailers were preloaded with ammunition, supplies and other stores (food, etc.). The supplies were then taken to a supply dump which was set up by the L.C.C.

The L.C.C. both parachuted and arrived by glider onto the DZ/LZ. Their job was to collect and distribute supplies to the various troops. The L.C.C. also collected supplies that were later dropped by R.A.S.C. Air Dispatch on resupply runs. The L.C.C. were also responsible for setting up security parties for the DZ/LZ, and like the Glider Pilots the L.C.C. defended the DZ/LZ and kept the area clear from the enemy.

An R.A.S.C. Corporal goes over his record book accounting for those stores issued and those stacked in the ammo/supply dump. Note the issue Driver's goggles, which were commonly seen on jeep drivers. This driver is in shirt sleeve order and wears a KD (Khaki Drill) shirt as he is a veteran of Sicily where tropical kit was issued. Note the issue trouser braces made of cotton with leather tabs, and his cap badge through the goggle lens.

A better view of the trooper as he does his books. On the chair rests his Austerity Pattern blouse. The large crates in front of the table are Rifle ammunition and artillery shells.

Each man wore a pair of two identity discs around the neck as per regulation. The tents are typical of what the Divisional troops might have at their Rear Area Command Post or Supply Dump.

For King and Country: British Airborne Uniforms, Insignia & Equipment in World War II

Left: An R.A.S.C. Supply Dump showing petrol cans, water cans, and trainers filled with weapons and ammunition. Right: Another view of the R.A.S.C. Supply Dump. To the right is the seat and tail of the Welbike and behind it is the BSA. Sleeping bags are laid out for a short kip (cat nap). It must be Tea Time as all are elsewhere, probably huddled around the Tommy Cooker (small stove).

An R.A.S.C. L.C.C. Driver pulls up to a just landed drop container. He's on the scene to gather supplies and organize a supply dump. Note the Mickey Mouse camouflage painted on the Airborne jeep. In the background a Para has just landed and is now on his way to lend a hand to his mate whose had a hard landing.

Chapter Three: Corps and Administrative Departments

Note the supplies in the trailer – this trooper has been busy collecting supplies as they land. To the right is a Para after a hard landing as he is on his back. A Para was taught to do a PLF, the Parachute Landing Fall. As the ground approaches, he prepares himself for the PLF with feet and knees together. Upon touching the ground he will then tuck and roll to the right or left, his hips absorbing the shock. He will then roll over landing on his side. He will then collapse his chute as soon as possible otherwise the still inflated chute can drag him across the DZ. In this case the trooper came down too fast. His chute may not have fully deployed thus his decent is that much faster than required for a safe landing.

An R.A.S.C. man on a BSA bicycle lends a hand with the containers and collapses chutes as they cannot remain on the DZ lest they become entangled with those troops and supplies yet to come.

For King and Country: British Airborne Uniforms, Insignia & Equipment in World War II

A parachute will remain deployed (inflated) until the air is 'spilled' out of it. These chutes can be used again or made into make shift tents. The Dutch also made good use of them in making silk under garments, curtains, and bedding.

All lend a hand in loading the trailer. On the left is a 1st pattern smock. Note how the tail is left to hang down as their are no rear securing snaps. Hanging from his waist is the issue clasp knife and lanyard. An R.A. Gunner makes room for the PIAT. The R.A. had a common practice of hanging the shell dressing from the Denison shoulder strap. To his right is the Airborne Sleeping bag – seen here is the water-proofed bottom. The Driver also wears a 1st pattern smock which has faded from wear and exposure. Also shown here is the box-shaped jeep trailer. Note the wheels of the trailer are not underneath yet rather outside of the body allowing for more storage room.

An R.A.S.C. Captain 'supervises' his troops in loading the trailer. Around his neck is the Webley revolver lanyard. He wears an other ranks badge upon his beret. His wearing of the Body Armour is an exception as it was mainly issued to the Recce Squadron and to the Poles. To his left is the butt plate of the MK V Sten. Also of interest is the latch to where cleaning tools are kept.

Chapter Three: Corps and Administrative Departments

The LZ

The LZ (pronounced L-zed) was the landing zone for gliders of the Airlanding Brigade. Here we see an R.A.S.C. L.C.C. gathering supplies off the LZ. These supplies will be taken to a supply dump, which the R.A.S.C. will set up. Their job was never done. I think of all the troops, these men were some of the hardest working and most unappreciated by today's military enthusiasts.

A supply dump. Tactically the troopers advance to where the dump will be established. Each trooper advances cautiously watching his side for the enemy as they have by now seen the transports and descending chutes. Underneath the Officers arm is the issue P-37 map case. The Bren Gunner and jeep follow behind. If caught in the open the Bren will go to ground and provide suppressing fire for those in front who will flank the enemy.

Of interest is how the Bren is carried by its wooden handle. It can also be slung by its web shoulder strap and carried in that fashion. Note how the jeep's front bumper (fenders) are cut off and the spare tire secured in front of the radiator. Below this can be seen the pick axe handle. This tool will be secured to the side of the jeep and can be easily slipped over the top of the tool handle.

The jeep is well loaded to full capacity. Here is a better look at the trailer wheel assembly. Basically it's an open box on wheels – perfect for the Airborne being able to carry everything needed from troops to kit.

53

For King and Country: British Airborne Uniforms, Insignia & Equipment in World War II

Left: Note the metal frame holder atop the jeep bumper (fender). This is to hold extra batteries as seen held in the P-37 satchel. Right: On this side as well there is another metal frame atop the bumper (fender) for holding the spare batteries. Note how the E Tool is carried on the Driver's webbing – this is its common position. Note that the handle can easily slip out and often did as there is no lanyard or securing side strap. The Drivers beret is tucked under the Denison shoulder strap.

Left: Note the dangling clasp knife of the Driver. Here can be seen the fading 'water color' effect of the Drivers 1st pattern smock compared to the Captain's late 1946 post-war variant's sharp contrasts and lines. Right: To the left of the Driver is a large rolled up camouflaged net commonly carried on the jeeps. This net could be thrown over the jeep thus hiding it from enemy aircraft.

Note the Driver wears his Ape tail between his legs, this was uncommon as most let them simply hang down.

54

Chapter Three: Corps and Administrative Departments

Note the smaller toggle worn by the Sten Gunner. This is about 1 to 2" shorter than the common issue toggle. Also of interest is the clear overlay of the map case. The earlier variant featured a hard board and did not have the clear overlay map pocket.

"Have a fag sir?" And the lads light up and take a bit of a breather. Note the Driver has cut his hand while collecting supplies. All in a day's work!

The cigarette box is the issue Players cigarettes, and was actually issued for troops on leave. There was a special cigarette tin that was to be carried in the field. There were two sizes of Players, one containing 5 and another containing 10 cigarettes. Players is simply a Brand name as other brand names were also issued.

Right: This is the British made Shell Dressing. Two of these were carried in the Parachutist Trouser twin rear pockets. This is the dressing/bandage seen hanging from many a Denison worn by the 1st Airborne. It was mainly the R.A. troops who wore the dressing in this manner.

55

For King and Country: British Airborne Uniforms, Insignia & Equipment in World War II

Right: As often happened due to bad weather or flak, troops were mis-dropped or their gliders thrown off course. Note the small open web case atop the jeep in front of the driver – this is the P-37 Binoculars case. The map case is the special Airborne Tri-folding variant. The jump wing on the Captain's smock is the most common variant and a nice example at that. The backing is serge like that of the BD yet heavier.

Right: A better look at the camouflage net draped across the hood. Note how every inch of space has been well utilized. The Bren Gunner is sitting on top of the folding stretcher. On the side of the jeep are the axe and general purpose shovel (spade).

Below: The Sten Gunner is sitting on top of a Bergen Rucksack, folding Bicycle, ammo boxes and all sorts. Of interest is the rugged construction of the trailer. Not the small metal chest with web carry strap underneath the Bergen – this is a Bren Ammo chest.

56

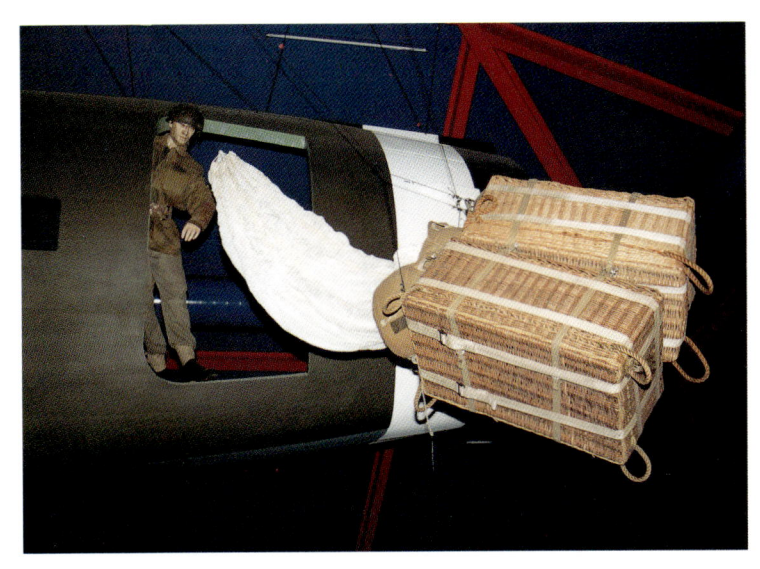

Chapter Three: Corps and Administrative Departments

Left: Two "daisy chained" panniers as they are dropped from the Dakota. A pannier was simply a very large wicker basket and held up to 350 pounds of supplies which was then dropped by parachute to the troops below. A dispatch crew was required to eject the panniers from the aircraft, either manually or by a special roller conveyor. Each pannier was about 36" x 20" x 16" deep and in halves. The bottom half fits into the top half which then forms the lid. The pannier can be expanded up to a depth of 30". Two leather straps hold the two halves together. Each strap has a metal buckle which is fitted with a quick release device which in aid of those recipients below to get it open quickly as they may and often were under fire. To slide the pannier on the aircraft floor or on the roller conveyer a sheet of plywood or battens were secured to the bottom of the pannier. Empty the pannier weighs about 50 pounds. The parachute is attached by a web harness. Panniers can be dropped separately or in pairs as seen here. This double method is called "daisy chaining." Right: Each pannier and parachute is complete in itself. When dropped as a "daisy chain" the panniers are then strapped together one on top of the other with four No.8 cords. These are tied between the leather straps and panniers. Each strap has a self-adjusting buckle and D rings for the attachment of the parachute bag. The plywood base or battens are only attached to the bottom pannier (not shown on this display). Note the rope loops for carrying and moving the Panniers which are found on each corner of the pannier.

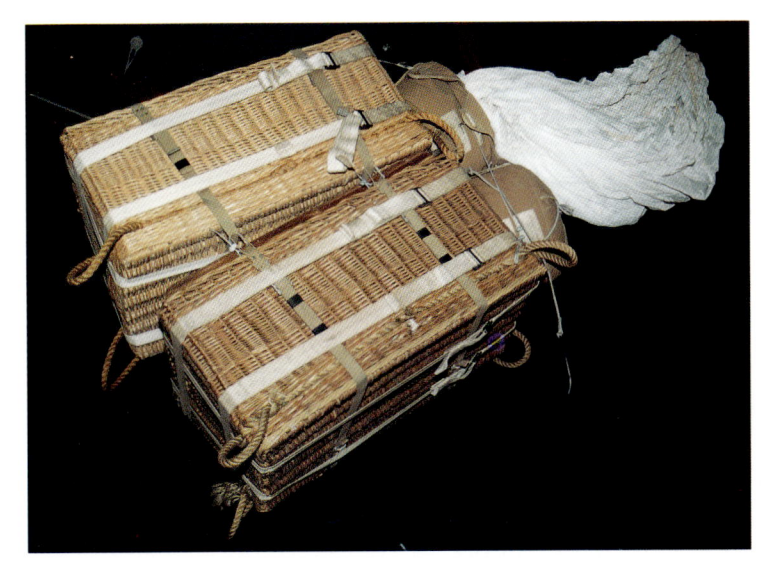

Left: The static line of the bottom parachute is attached to the nearest leather strap on the top pannier, yet leaving a bit of slack. A 10 foot extension cord is attached to one end of the static line of the top parachute. On the other end is a snap hook for attachment to the aircraft strong point. Shown here is the parachute just as it is beginning to deploy – the chute will open safely after the aircraft has flown on. Right: When the panniers are dropped the top (pannier) parachute opens first, and as the fall is checked by the deployment of the chute the bottom pannier will break away. It will do so to the limit of it's static line, then the bottom panniers chute will open, thus the panniers then fall separately.

Chapter IV

The Glider Pilot Regiment

Glider Pilots were trained to be the 'Ultimate Soldier'. Training and qualification requirements far surpassed that of the Parachute Regiment. Pilots were first taught to operate all weapons and to more than hold their own on the battlefield. After their combat course the trainees were then taught to fly. Prospective Glider Pilots could enter the training course at the rank of Corporal. Upon the successful completion of their training and elementary flying course they were promoted to the rank of Sergeant.

The Glider Pilot Regiment was an all volunteer outfit with members coming from the Army. RAF pilots were in surplus in the later years of the war and were also trained to fly gliders. A glider pilot was first taught to fly a Tigermoth which was a single-engine biplane trainer. After about six months traing in the Tigermoth, he would then be converted to gliders and learn to fly each variant of glider.

Note the wrist tabs on the smock. This pilot has a Blood Stick. This is a regimental cane with the regimental design decorating the tip in sterling or brass. It is called the Blood Stick as many a trooper was thumped over the head with it for not paying attention, and or talking out of turn.

Left: A 1st Pilot from B Squadron. This Captain wears the Army Air Corps badge on his beret. Around his neck he wears a self-made scarf. This is U.S. parachute material. The U.S. Troopers often cut up spare parachutes to make scarves and was a common practice for Normandy. This pilot has been to Normandy, survived Arnhem and gone to the Rhine Crossing. He was given this scarf by members of the U.S. 17th Airborne Division. Of interest is the pristine color of his 2nd pattern smock – note how his rank is sewn to the Denison shoulder strap, as was done by regulation.

58

Chapter Four: The Glider Pilot Regiment

Those Who Served
Major Gerrard Roland Millar, B Squadron, Glider Pilot Regiment

I was commissioned in the Royal Inniskilling Fusiliers. Our company was posted at RAF St. Eval as ground defense. It was here that I first heard of a new unit called the Glider Regiment. I then applied and joined. On my first solo flight I had a bad prang (accident) which slowed me down for the rest of my life. It was pitch black night, coming in to land, I had a gentle bounce which told me all was well. Then all hell happened! The Hotspur's nose disintegrated as it hit the ground. The wooden back support hit me in the region of the first and second lumbar and broke itself in the process. I finished up in great pain lying on the grass and bits of cockpit floor. I do not know what happened that night and as the night was so dark, nobody saw the accident. The MO (Medical Officer) there was at the time quite a queer fish. I was not X-rayed. Good job I wasn't as I would have been grounded! I could wriggle my toes so therefore was ok. Remarking that I was in great pain, he tied my legs at the end of the stretcher so that I could not move. The magic cure worked and I was back in the air again in ten days.

I then joined Major Ian Toler's 'B' Squadron at Stoney Cross. I was one of the unlucky ones who missed D-Day but was with Jimmie Barclay for Market Garden 297 Sqn (Squadron) was our tug aircraft. When NE of Tilberg we flew into a violent cumulonimbus cloud and were tossed all over the place, visibility nil. At one stage the slack rope yoke was resting on the cockpit roof then it slowly moved back behind and out of sight. After what seemed an age, we were tossed up out at the top (of the cloud). The Tug aircraft emerged from underneath, however, the port yoke was around the wing tip. That broke when the slack was taken up. We flew on the starboard yoke for a few seconds, then one strand was cut by flak. The remaining two just frayed to nothing. We had to land near cover and go to ground. That evening the gallant Dutch resistance contacted us and took charge. Six weeks later we liberated Boxtel and the next morning met the Seaforths of the 51st Division.

Next 'Varsity' from Earls Colne flying with Sgt. Husbands, my Second Pilot. My main memory of Varsity is that of thick smoke. We could not see the ground. I hit a row of low tension wires which came away. Apart from a damaged nose we landed ok. We had a hell of a time unloading the gun, jeep and trailer under fire. I was wounded in the hand. The next day Sgt. Husbands went around to identify the dead. 100 Glider Pilots had been killed. At the top of our LZ three Horsas were fairly close to each other. Two had belly landed, the other was perfect. Going to it first we saw both pilots strapped dead in their seats. Entering we saw all 26 passengers sitting dead in their seats. No blood or any sign of damage. They were alive one second and dead the next. In the cockpit I found a good friend of mine, a fellow Irishman, Willie Welply from Limerick (he was the pilot). There apparently unharmed both pilots sat there dead. The other two gliders were also carrying troops. All looked alive but were dead. I discussed the situation with others who had seen the "dead/live gliders." Was it gas? If blast, (flak) why were the gliders in such good condition? So be it.

On our return we went to Keevil. Major Toler told me we needed a training camp before going to the Far East. "Go off, find a place, set it up and let me know when to send down crews" The result was Watermouth Camp, Ilfracombe (a seaside resort in Wales) and I can assure you that damn little training was done down there! Around this time most of us converted on to WACOs.

In a glider, there are two pilots, the 1st and the 2nd. The 1st pilot is the main pilot of the aircraft. To become the 1st pilot he had to pass the flying courses with high marks (grades), while the 2nd pilot passed the same course with lower marks. The 2nd pilot was what we call today, a co-pilot, and his duty was to assist the 1st pilot, for example by spelling him (taking over the controls while in flight and giving the 1st pilot a break) as flying a glider was quite strenuous.

An Officer was not necessarily a first pilot. If the officer did not complete the higher degree of training required for a first pilots position he graduate as a 2nd Pilot. This did happen.

Glider Pilots were originally part of the Army Air Corps until 1942 when they became their own regiment, the Glider Pilot Regiment. They did not have their own regimental badge though until 1947.

This is the 1st Pilots brevet (wing). It was worn both on the smock and also on the BD.

Transport

The Glider Pilot Regiment provided the pilots who flew troop transports, such as:

Hotspurs were designed to carry eight troops. This was the very first glider introduced and used. Glider Pilots trained with these after their first initial flights in the Tigermoth, a biplane trainer.

Hadrians, which were the American made WACOs which could carry 15 troops or one jeep with trailer or one jeep with 75MM howitzer or one 6 pounder anti-tank gun.

Horsas were designed to carry 29 troops with their equipment or special equipment and vehicles such as a jeep and 6 Pounder.

Hamlicars were designed to carry heavy weapons as one 17 pounder anti-tank gun with Morris 15 cwt (truck), or two Bren Gun Carriers, three Rota trailers, two Dingo Scout Cars, 48 equipment/supply panniers, a 25 Pound field gun with tractor, a self propelled Bofors gun, one Jeep and Bren Gun Carrier, a Bren Carrier with 3" mortar and 8 Motorcycles, one Tetrarch light tank, one Locust T.9 light tank, or a Bailey pontoon bridge.

A very rare and unusual smock. Towards the end of the war there was a surplus of RAF pilots – these surplus pilots were retrained to fly gliders. The pilot's wings are RAF as seen above the left breast pocket. On the right sleeve are Flight Sergeant's chevrons – they are identified as RAF chevrons as they are light blue on dark RAF blue serge. Note that they are worn only on one wing. This is a 1st pattern smock and has not been converted to full zip which is itself rare. Tell tale signs of the 1st pattern smock are the tapered sleeves, knit cuffs and lack of wrist tabs. Also, the base color is a very light tan/khaki which is seen mostly on 1942 dated smocks. This base color is uncommon compared to the ratio of those that are most commonly seen by the collector. (Courtesy Andy Birt collection)

Right: Close up of the Pilot's brevet and size label – the labels do wear out easily from light wear.

Chapter Four: The Glider Pilot Regiment

Left: The history of this smock is as follows: Les Kershaw was sent to Canada for training to learn how to fly multi-engine aircraft as a Sgt. Pilot – this was common during wartime as usually a pilot was none the less a junior officer. Just before the Normandy Invasion Les returned to England with his class of Sgt. Pilots. When they arrived they were told that there was no aircraft for them to fly. A week passed by and they were then put on parade. There they were told that they were to be trained to fly. Les's reply was, "excuse me but we can already fly!" The reply came back, "You haven't flown gliders!" From there they were converted to gliders. Kershaw wore this Denison when he flew his glider over the Rhine for Operation Varsity. Of interest is the name tag above the right breast pocket. Right: A better look at the RAF rank chevron. Note the Squadron tapes sewn to the shoulder straps (Glider Pilots were put in squadrons and not Battalions like the Paratroops or Glider Infantry). These are unique for Varsity as Kershaw only wore this smock for that operation.

Left: A full length view of the smock. Note its excellent condition, again as it was worn only for one operation. Right: A better look at the RAF Sergeant's chevron and squadron tape. Note that it is crudely cut and sewn to the strap.

Chapter V

Issue Clothing

Under Garments

Drawers, wool
These were standard issue of knee length with a cotton 6" waist band with either bone or plastic buttons. Commonly each pair had its size and date stamped on the outside waist band.

Drawers, wool, winter
Full length tops and bottoms. The bottoms have a 6" waist band and plastic buttons. The top have a rounded crew neck collar, much like that of a sweat shirt. Commonly each pair had its size and date stamped on the outside waist band.

Issue under pants (the Drawers), Wool. Common variants feature black ink stamping on the rear waistband, consisting of manufacturer, date, size and or size and size information, and broad arrow. This pair features a small size label on the inside waist band.

Size label of the Drawers wool. Note the plastic button which is the same found on early Battledress Serge 1940 Pattern blouses and trousers.

Left: Here is seen the plastic Recce badge proudly adorning the beret. Also of interest is the Collarless shirt non-collar.

Chapter Five: Issue Clothing

Collarless Shirt

Made of angola (flannel) wool with a khaki twill 1/2" banded collar for other ranks with a single button to close the collar and two more buttons on the front placket. This was the issue undershirt, looking much like a large night shirt. Variants are found to have either small plastic, vegetable or metal dish buttons (early war). Designed as a pullover garment, opening to the lower breast. Some special Officers private purchase variants were cut like an every day long sleeve shirt, opening completely. Officers shirts were often made of a finer wool. As Officers wore ties, they were issued with a button in collar, attached with two small studs which were made of bone, metal or plastic. As far as I know these shirts ranged in sizes from 1 to 4, 4 being the largest.

Other ranks were authorized to wear ties in November of 1944, yet only for "walkin' out" and not in combat. Note that there were limited exceptions, such as No.4 Commando C troop were allowed to wear them in the early years of the war as quoted by Bren Gunner George Jones. This was a special privilege. Most photographic evidence supports this.

Above: Note the color variation of the non collar this being of a darker twill. Also the buttons are different as each manufacturer will use different finishing supplies of what is available. The cord around the waist is a jackknife lanyard – the knife will be carried inside his pocket. Also of interest is the plastic cap badge upon his beret. Below: The button-in collar. Note the stamping: manufacture, broad arrow, date, and to the right a size number.

Above: Note the variation of Angola wool used in the manufacture of the shirt. Also note the variation in buttons used on this shirt compared to those featured here. This Signalman will surely be written up for the tea biscuit lodged in his waist belt. He is wearing the Parachutist trousers. As can be seen by his age many of the new replacements in 1st Airborne were very young. Below: Note the cotton braces and how with the adjustment of the buckle it can be lengthened or shortened. This Para is wearing a button in collar – this was an exception for Other Ranks. Of interest is how the cap badge hangs off the beret – a small slit and the badge is slid through. Earlier badges were poked through with two eyelet holes and a brass pin was run through on the inside to secure it.

63

For King and Country: British Airborne Uniforms, Insignia & Equipment in World War II

Braces, trouser (suspenders)

As belts were not of common issue braces were worn to hold up the trousers. They were made of white cotton non elastic and most had leather button tabs on each end. Some pairs did feature cotton tabs, yet this is rare. Each pair adjusted for comfort. I know of three sizes, small, normal, and large. Others are marked with a number such as 36, or 38. It is pure speculation as to what these numbers indicate.

String vest

Loosely knitted of tightly woven cord (string) and designed to be worn either under the shirt and against the skin or over the shirt to keep dry and either warm and or cool-thus maintaining body temperature. There are two patterns. The first is without a cloth gathered shoulder band. This is known as the 'Escape' pattern as it could be unraveled into a long cord since it was all just on piece. A second pattern with a gathered shoulder which was more comfortable to wear as with all the gear and weight on the shoulder the skin would be worn and irritated. Vests came in four sizes, sz 1 small, sz 2 medium, sz 3 large and sz 4 XL. The size label was looped around the bottom edge of the vest. This piece of clothing was manufactured well into the post-war years.

A Corporal conducts an impromptu weapons drill by explaining the wonders of the MK II Sten. Also seen here are the issue cotton trouser braces with leather tabs which came in three sizes: small, normal and large.

A Para cools off in the courtyard of a small French farmhouse. This is one of two variants of the String Vest. It was designed to be worn as a first layer yet was also at times worn over the Collarless Shirt. The vest works quite well in keeping the body warm in colder climates, and also absorbs sweat in hotter climates. Note the gathered shoulder. This is the difference between the two vests as the other does not have a gathered shoulder.

Chapter Five: Issue Clothing

This is the manufacturer's label of the String Vest simply sewn over a loop in the vest.

A soldier's first love ... digging slit trenches. He is using the General Purpose Shovel issued one per section of ten men. Here he wears the vest as a first layer. Note the ladder laced boots as per regulation.

Flip side of the label. To the author's knowledge there were 4 sizes, with 4 being the largest. Note the Broad Arrow to the right of the date and size stamp. Most vests in this 'gathered' shoulder pattern seem to be dated 1943 and 1944. As most of the ungathered 'Escape' patterns seem to be dated 1945.

To the left is the 'Escape' variant and on the right the gathered shoulder variant. Note that both variants seem to be dated from 1943 to 1945. Therefore it is not a matter of which came first. More investigations are in order. (Note that the entitled 'Escape' variant may simply be a term introduced by collectors and not official.)

Battledress

Battledress was modeled after pre-war ski wear that was high-waisted and loosely fitting trousers and jacket which allowed for layers to be worn underneath. The British Army was in the process of revamping its general issue uniform and equipment as those currently in issue had been in use since World War I.

Battledress was manufactured by each of the Commonwealth countries yet not all were of the same quality. Canadian made was by far the best and quite sought after as it was made of a superior wool that was comfortable to wear compared to the scratchy wool used by the other countries. British wool ranked second in quality, from there it went downhill from Australian, South African, and Indian. Of interest is that the airborne troops were also issued with U.S. made BD. For example, the Hartenstein Airborne Museum has a blouse from the 21st IPC which is of U.S. manufacture and was part of the Lend-Lease Program. The U.S. manufactured British Battledress and the British made uniforms for the British.

British Battledress was made of a khaki serge wool. Most soldiers were issued with two sets. One set for Parade, duties (work details), and for combat. A second set was reserved specially for nights out on the town, on leave or special events as a British soldier had to look smart (sharp!). This was know as a soldier's 'Best Dress' or his 'Walking Outs'. This best dress set was kept in the soldiers kit bag and not taken into combat as paratroops are limited for space and did not have the 2nd Echelon units to bring up their personal effects like some of the fortunate line units did.

Battledress was made to be worn buttoned up and closed at the collar. Officers were by regulation to wear a tie so they wore their blouses open at the collar. There were common alterations made for the officers such as facing the inside lapels and neck with serge. Some blouses were specially tailored to remain open. This was not commonly seen in combat but rather on Parade and for special events or on leave.

Battledress had 18 basic sizes – numbered size 1 to size 18. There were special sizes labeled *extra small*, *small* and *oversize*. The basic 18 sizes were grouped in threes according to height. Size 1-3 had the same height and leg size, yet each had different breast and waist sizes. Size 1 had the smallest waist and shortest leg where as size 3 had the largest waist and longest leg. The next set of three were size 4-6. Size 4 would be of a smaller waist size than 3 but was for a greater height. Size labels were made of a stiff white cotton or paper, each with the details stamped upon them. On each label you'll find the type of garment, for example Battledress Blouse Serge, then the size number, sizes which included height, breast and waist, manufacturer, date of manufacture and the broad arrow. A date of issue was stamped onto the label in ink which did not fair well with age as these days you'll see a blur of blue or black ink. Labels were sewn to the inside of the blouse internal breast pocket and on the outside rear of the trousers. As these labels were made of thin cotton and paper most have long since cracked and or deteriorated.

The relation between sizes made and numbers produced were more smalls and mediums, and fewer of the larges and extra larges as the English were not large people. This is a fact that few collectors and even fewer reenactors realize.

Size numbers were also stamped in ink on the internal breast pocket and to the inside waist. Also found stamped onto the blouses and trousers as the size numbers were was a WD (War Department) and broad arrow stamp with a manufacturer's code.

Here a trainee awaits his turn to jump from the balloon. He wears the Battledress Serge 1940 Pattern blouse. Note the pleated breast pockets and revolving shank shoulder strap buttons. The blouses front and pocket buttons are made of either brass or metal and concealed. Twin eye hooks secure the collar. The collar is lined with a khaki twill which is the main difference and improvement of this pattern. The internal breast pockets are also lined with this same twill. He wears the AIRBORNE shoulder title, and below this the embroidered Pegasus flash. He also wears the Bungey training helmet. This variant has no grommets or slits made in its side flaps. Note the clasp knife lanyard around his waist.

Right: Size label for the AIRBORNE Battledress Serge 1940 Pattern blouse. Many collectors wrongly call this the Pattern 37 blouse. This is a typical size label, featuring: garment type, size number, size information, manufacturer, date of manufacture and broad arrow.

Chapter Five: Issue Clothing

Battledress Serge

Battledress Serge was first issued in late 1938. Commonly known by collectors as pattern 37 as this was the year that the Army approved it for use, as well as the year in which the new webbing was approved. Though there is no such designation nor label.

Battledress Serge Blouses featured an unlined collar, a simple metal buckle on the waist, two pleated breast pockets, two internal breast pockets made of twill, two shoulder straps with revolving shank buttons, and a button up front with twin hooks and eyes to secure the collar. All buttons were brass and were concealed with serge flaps. On the rear waistband were three button holes to which the trousers would fit through. In 1939 the breast pockets were raised 1/2" to allow easier access to the pockets while wearing the new P-37 webbing.

Battledress Serge Trousers featured a patch map pocket on the front left leg and a first field dressing on the upper front right leg pocket, and was to be removed only under orders from an officer (as it was the King's property), two side pockets lined with cotton twill, one rear hip pocket, five belt loops, and blousing tabs on the bottom of each leg to close the leg when worn with P-37 gaiters. The blousing tabs would be undone when worn without gaiters. The first field dressing had a single pleat and both the map and rear hip pockets had flaps. All buttons were brass and concealed. On the rear waist are three buttons to attach to the blouse. Buttons were also found on the inside waist for the use of braces (suspenders).

Battledress Serge 1940 Pattern

This is the second pattern of battledress and was very similar to the first yet with the following changes: in May of 1940 the collars were lined with cotton twill as the earlier collars were bare serge. This rubbed the neck raw making it most uncomfortable. In June of 1940 the trouser first field dressing pocket had a box pleat added in place of its original single pleat. In July of 1940 a new waist buckle was introduced which was made of steel and featured teeth to better secure the side waist belt. The first variant of buckle was a simple loop and became easily undone. These first buckles were still in production use until stocks ran out in 1942. The buttons were now of a galvanized metal, yet brass buttons were used until stocks ran out.

Another Battledress Serge 1940 pattern blouse. Note the steel buckle with metal teeth, as the first pattern blouse, the Battledress Serge, featured a simple brass or steel loop buckle and brass buttons.

The concealed steel button of the 2nd pattern blouse, Battledress Serge 1940 Pattern. Note the pleated breast pockets and concealed buttons, which are concealed by button flaps.

Battledress 1940 Pattern (*Austerity Pattern*)

Known to collectors as the 40 pattern it is actually the Austerity pattern – this was in aid of the war effort to save materials and production costs. Major changes were made on this last to be issued blouse of the war. The next pattern was introduced in 1946 and known as the 46 pattern. These changes were no pleats on the breast pockets, all of its button flaps were removed and the buttons left exposed, brass and metal buttons were replaced with first chocolate and later olive plastic buttons, the

Left: Shown here are the major differences between the Battledress Serge 1940 Pattern and the Austerity Pattern (3rd pattern of blouse manufactured) which was made in aid of the war effort. Material was saved by not pleating the pockets and deleting button flaps. All Austerity blouses feature the all steel waist buckle and plastic buttons. Plastic was also used to save money and materials. Collectors wrongly identify this blouse as the 1940 pattern.

buttons and button holes on the rear waist were decreased to two from the original three, soon after its first issue the twin internal breast pockets were replaced by just one.

The Austerity trousers were void of blousing tabs and belt loops. All buttons were now plastic. First chocolate and then olive. All button flaps were removed to expose the buttons. Revolving shank buttons were also replaced by plastic ones.

War time trousers of any pattern can be one of the hardest pieces to collect. Few have survived. Upon leaving the service, or as the ol' boys say "De-mobbing" (de-mobilization), a soldier was allowed to keep two pair of trousers. One such pair could be dyed and worn to work and or college for example.

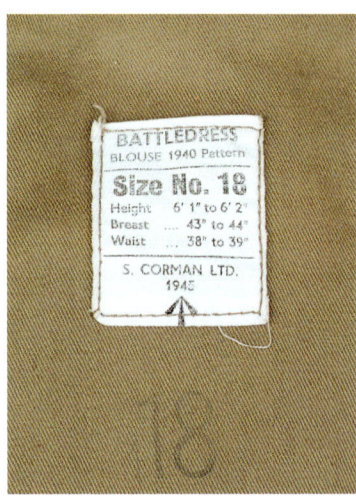

Left: The cuff and its exposed plastic button. Most later dated blouses seem to feature the olive buttons as seen here. Those dated earlier have chocolate colored buttons. The brass stripe is a Wound Stripe which was awarded for wounds received in combat. Note the Battledress Serge 1940 Pattern Trouser map pocket with its exposed button flap. Right: This is the size label much like that found in all blouses. Note that the configuration of the stamp from top to bottom: type of garment, size number, sizes of men who can wear the garment, manufacturer, date of manufacture, and Broad Arrow. Also, in a blue or red ink (usually) the date of issue will be stamped across this label – these issuing stamps do not fair well with age. Also note the ink '18' stamp below the label – most blouses will have these. There will also be a War Department (WD) Broad Arrow stamp on the blouse's internal pocket.

Shown here is a 'fake' Commando Austerity Pattern blouse. The collector should be aware that just because the patches are on a blouse that they are not necessarily native to it. This is rampant among elite blouses, mostly Airborne and Commando. Here are shown the very early white on black shoulder titles on a 1945 dated blouse. These titles were worn from 1940 until 1942 (and with exceptions until 1943), yet for them to be found on a 1945 dated blouse is most unusual. The Commando wore consistently red on black blue titles from 1944 until the end of the war. So why the early titles on a late blouse? The early jump badge and not jump wings, as the Commando wore the same wings as the Paras from 1944 to the end of the war. Another tell tale sign is that the shoulder titles were sewn on with nylon thread. Nylon thread was not issued in the House Wife sewing kits during the war. Not to mention that there is no fading, as if the titles had been there since the war, of the material underneath the titles – this should be darker than the body of the blouse.

An interesting pair of Austerity pattern trousers with brass buttons. Plastic buttons are found on the dressing and map pockets.

Chapter Five: Issue Clothing

Shown here is the size stamping on the twill lining – also note the WD and broad arrow stamp. It is very unusual that brass buttons are on a pair of trousers such as the Austerity pattern, possibly because the manufacturer was using materials that were on hand.

Here can be seen the single rear pocket and the twin buttons used for the attachment of the blouse. Of interest is that these buttons are also plastic, as brass was used on the button fly and trouser brace buttons found on the inside waist.

The size label on this pair of Austerity trousers. Just above the broad arrow can be seen the issue date stamping of Oct. 1942. Note how the label does not read "Battledress Serge", but rather Battledress Trousers 1940 Pattern. This is yet another break in patterns from the two earlier patterns and the last, Austerity (third) Pattern. The term Austerity is used for this pattern which was designed to save money and materials.

This is the Austerity Pattern Trouser First Field Dressing pocket and its First Field dressing. This pocket is secured by a plastic button as shown. The dressing cover once open reveals a pair of dressings, one of which can be seen in use on the soldiers arm. The other dressing is still inside the cover.

Above: Close up of the Austerity Patterns First Field Dressing pocket, empty. By regulation a dressing was to be carried within the pocket.

Left: Rear view of the Austerity Pattern trousers. Note the two buttons on each hip. These are for the attachment of the blouse which has two button holes corresponding to these buttons. One rear hip pocket is featured on the trousers and as can be seen the cotton trouser braces with leather tabs which attach from the inside of the trousers.

69

Trousers, Parachutist

Specially designed for the Paras to offer a better suited uniform for their special role, and first issued in 1942. Like the P-40 all buttons are exposed and of plastic. The trousers Featured an extended bellows map pocket lined with chamois for extra strength which could be used to carry mills bombs, Sten magazines, and what ever else one could stuff in there! This pocket had two small 1/4" to 1/2" metal studs and one plastic button in the middle of the flap. Two side pockets were also lined with chamois. A first field dressing pocket was found on the right front leg and two Shell Dressing pockets on the rear in place of hip pockets. Again, these dressings were to be only used upon the order of an officer. My Uncle thought twice before opening his at Arnhem as such disciplines were highly enforced by the British Army. Also featured was a built in FS dagger pocket, the hilt of the Fairbain Sikes dagger would rest upon its opening and only the hilt and handle would show. A plastic button would secure the top tip of the leather sheath to the upper thigh.

Some variants had small metal eye hooks or brass snaps to secure the pocket opening while the wearer was in mid-air. No belt loops came on these trousers so therefore braces were worn. Not all Paras were issued with these special trousers. All patterns of trousers were worn in the field at the same time by the same unit.

Note the bellows (extended) map pocket of the Parachutist Trousers. On the right front leg is again the First Field Dressing pocket. There are no blousing tabs on the legs. (Courtesy Michel De Trez)

Note that there are no belt loops on either pair, and how close the Shell Dressing pockets are to each other. Most reproductions will have these pockets spread widely apart (so few original samples are available that many collectors have never seen a real pair). Note the differently colored buttons used. On the left a darker chocolate and on the right the very common olive. Also of interest is the varying colors of wool between the two pairs. A collector must make note that not all was the same and matched perfectly. One last observation is the difference in the sizes of the labels.

Note the differences between the 1943 pair on the left and the 1945 pair on the right. On the 1943 are steel buttons on the fly, and plastic buttons on the 1945. Also note the difference in color of the cotton twill lining. This is yet another manufacturer variation.

Detail of the bellows pocket.

Right: On the 1943 pair is a chamois lined bellows pocket, and on the 1945 pair a cotton twill reinforced bellows pocket. The pockets were reinforced to strengthen the pocket, allowing the wearer to carry Sten magazines, grenades, etc. Should the pocket not be reinforced it could blow out and its contents spill all over.

Chapter Five: Issue Clothing

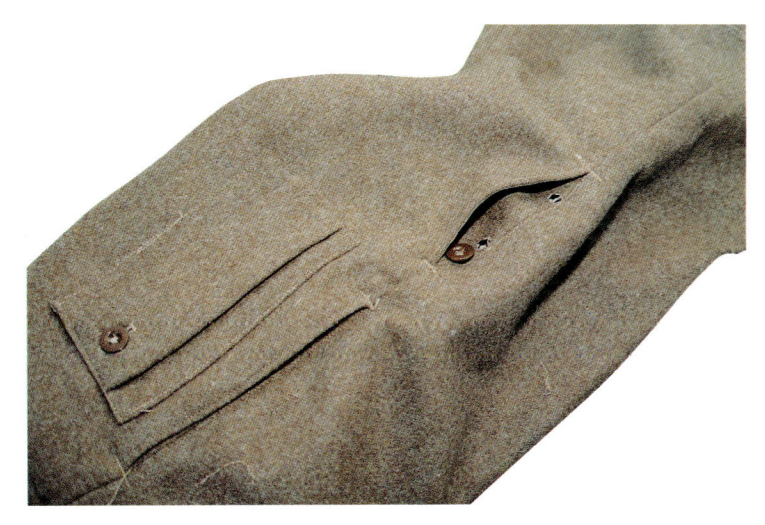

Left: Note that the corner is not sewn to the trouser leg but is folded and pleated making the pocket hang outward from the leg. Right: Shown here is the front Dressing pocket and also the FS Dagger pocket. The button was designed to fit through the slit in the leather dagger sheath which held it in place – this button could only be used if the pocket was open, and note that there are twin snaps which close and secure the pocket. Some variants are known to also have a hook and eyelet which also secure the pocket and dagger.

Left: Closer look at the FS Dagger pocket. Right: Trouser label: Garment name, size number, size of man who can wear it, manufacturer, Broad Arrow, and date.

Left: Closer look at the Shell Dressing pockets. Note that the odd colored button has either been replaced or was actually made the way it is shown here. Right: Close up of the chamois lined bellows pocket.

Left: Note the steel buttons which are for the attachment of braces. Also note the War Department stamp. In the middle is a Broad Arrow and below a manufacturer's code. Both hip pockets are made of cotton twill. Right: Just visible is the faded issue stamp just below the manufacturer's name. This pair was issued on February 10th – the year is not readable.

Left: On further inspection note that the middle waist button is sewn on with a different thread. This may have been added by its previous owner so as to be worn with a Battledress Serge or BD Serge 1940 Pattern blouse, which had three button holes to the Austerity's two rear button holes. Right: The cotton twill lined 1945 bellows pocket. Note the inverted middle of the pocket.

Left: The brass bellows pocket snap. To the author's knowledge this snap was only used on these trousers. Right: And the reverse underside of the bellows pocket flap – the manufacturer's name and patent number is just visible.

Chapter Five: Issue Clothing

Smock, Airborne Troops

This was the first piece of specialist clothing made for the British Airborne yet it was copied from the German Fallschirmjäger one piece 'Bone Bag'. The British variant was made of an olive gabardine denim-like material. It featured two short, just above the knee, step in legs with a concealed full length frontal zipper. On each waist side was a slit pocket which allowed the wearer to access his Battledress underneath. These pockets were each secured by a zipper. Long sleeves with outer and inner cuffs of which the outer had an extra snap making it adjustable. The inner cuffs were elasticized to retain warmth. Each outer leg was also adjustable with an outside snap. Three sets of closure snaps were found one each above the lower abdomen, top of the breast and the neck. Also found was a stand or fall collar. This smock was worn for the raids on Monte Vulture and Bruneval and was replaced by the camouflaged 1st pattern Denison in 1942.

Smock, Denison, Airborne Troops

Designed by Major Denison and introduced in 1942 it was made to be worn over the soldiers Battledress and under the denim Jacket Parachutist. There are two common patterns that were issued and worn during the war. These were the 1st and 2nd patterns. There are different variations of each pattern, for example the Full zip, Experimental and Snipers variants.

1st Pattern

1st pattern smocks were hand painted with large mop like brushes using various dye lots for camouflage coloring. The base material was an 8 ounce cotton twill. The idea of using the dyes was for the colors to diffuse upon its getting initially wet, thus the scheme would meld together and no sharp edges and or lines would be detected by an enemy. Once the smock was "broken in" the wearer would disappear into the surrounding terrain. Each smock differs from the next, and each has its own character due to the method of application of camouflaged colored dyes.

First pattern smocks featured a half steel zipper going to the base of the chest, two internal pockets, four external pockets which were secured with brass snaps which were made by the Newey Company. On the back was an "Ape Tail" which hung down while not between the legs as there were no snaps in the rear. (This is one version of how the Paras got their name 'Red Devils'. In North Africa the Paras were fighting in an area that was covered with red clay. The Germans saw these mad men coming at them caked with red dust and mud and tails flying behind them as they ran! Thus the name red devils! The more commonly known story is that they got their name from the Germans who remembered their special head dress, this being their red berets.)

The base color of the first pattern smocks was that of a yellowish sand with overlaying chocolate browns and pea greens. This base color was to best suit the wearer for the North African and Italian theaters. The collar was lined with angola wool. The sleeves were larger at the arm and elbow and then tapered to the wrist and had wool cuff sewn to them. Two shoulder straps were secured by the same plastic buttons found on the Battledress. Each smock would have a size label sewn to the inside lower waist. These size labels did not fair well with age and are the first thing to deteriorate with age.

It is thought by collectors that the tail (Ape Tail) was worn between the legs during decent to prevent the smock from blowing over the wearers head. But if you stop and think that with the Jacket Parachutist over the Denison and then on top of that the parachute harness the Denison is not going anywhere! I share the opinion that the tail added so that when a trooper crawled backwards the smock would not fold over his body or get tangled in the brush. When the tail was worn between the legs it was secured by a series of brass snaps manufactured by Newey. There were three rows of two snaps. This was to accommodate for the size and comfort of the wearer. Two waist adjuster snaps are also found, one on each side which can be worn either snapped or unsnapped. Newey snaps are featured on both the 1st and 2nd pattern smocks and were the only company to manufacture them during the war.

Here two Polish trainees enjoy a nice cup of tea from the YMCA Tea Van. Note the Quick Release box and harnesses. The trooper on the left wears the early 'Step In' smock. Note that it is full length and long sleeved. Both troopers wear a variant of the Bungey Training Helmet which has brass grommets featured on the flaps. These two were identified as Poles from the Hartenstein Museum otherwise there would be no way of telling by looking at the kit.

An Officer covers the troops as he and his men head back from a patrol in enemy held territory. Here the author wears a 1946 Denison that has had its wrists shortened and hose tops added. Note the P-37 Map Case under the left shoulder and Shell Dressing hanging from the shoulder strap. Also note that the holster is attached by a pair of P-37 brace attachments, and also the MK V Sten.

For King and Country: British Airborne Uniforms, Insignia & Equipment in World War II

Here is a 1st pattern smock – note the faded 'water color' effect, tapered sleeves and knit cuffs. This smock has been converted to a full zip – the zipper has been taken from the Jacket Parachutist. Also of interest is the camouflaged face veil worn as a scarf, as designed. Besides the full zip conversion this smock is perfect for a North African or Sicilian theme. The Para is armed with a No.4 Rifle.

One can just imagine how this smock was hand painted with large brush and dyes and of interest the direction of the brush strokes-one way, up and down. Here can be seen the lack of rear snaps with just the plain bare body – this is one of the main difference between a 1st and 2nd pattern smock.

Native to the smock, this is this common jump wing found on the right shoulder of both the Denison and BD blouse. The backing is thick serge wool. How to tell a WWII wing from a post-war is very easy – note how the feather threads of the wing are horizontal, this is a WWII wing. Post-War wing feathers are at an angle.

Chapter Five: Issue Clothing

Low on ammunition and lower on food this Para scans the skies for a hopeful supply drop. Of interest is the manner in which the boot laces are tied. This is by regulation and is called 'ladder' lacing. Note how the sleeves are looser at the arms and tapered to the wrist – this is the second feature and difference between a 1st and 2nd pattern smock. No wrist tabs are found on 1st pattern smocks. The cap badge is the plastic variant and a rare badge at that, with only 118,456 made.

Note the zipper and zipper pull tab. This makes it easier to open and close the smock rather than pinching the small zipper without it. Again note that this is the zipper from the Jacket Parachutist. Another dead give away is the all denim material of the pull tab. Also note the collar and how the paint/dye has spattered as the brush ended a stroke on the material.

Here is the waist closure snap. These allow the smock to be tightened around the upper thigh – it can also be undone to loosen. Note the contrast in colors used from the front body to the pocket and the faded rear.

Here the smock is turned inside out. The size labels are sewn 99% of the time in this place. They do wear easily however and are the first thing to age on the smock. Note the twin internal pockets – these are found on all war time smocks. Where this was once a half zip right where the top of the lower pockets, can be seen the color of the middle zip flap change. This is a separate piece of material and is in fact from the Jacket Parachutist.

75

For King and Country: British Airborne Uniforms, Insignia & Equipment in World War II

Left: These are the air ventilation holes found on each smock, under each arm. Note this is inside out. Right: This is the War Department Broad Arrow stamp. The numbers are a manufacturer's code. This stamp is found on most, if not all, British equipment and clothing and is the Army's proof stamp. Each piece is stamped after it is inspected.

Left: Shown here is a 1st pattern smock. Note the plastic cap badge on his beret. He's hidden his webbing under a tree and gone out to forage for food. Of interest are how low the breast pockets are. Right: Note the manner in which the paint/dye was applied on this smock as if it were spilt all over the front. There is little rhyme or reason to the 1st patterns camouflage schemes. The three rows of brass snaps in the front are for the attachment of the 'Ape tail'. This is worn between the legs during descent yet its purpose is such that when the soldier is crawling backwards the smock will not become entangled and/or fold back over his body. The purpose of the tail worn between the legs was originally thought to prevent the smock from flying up while in descent. Though with the parachute harness on the Denison is going nowhere. Not to mention if the Jacket Parachutist is worn over the Denison as well. There are few photographs showing the Paras wearing them between their legs in combat. This may be because it was uncomfortable to wear the smock for long periods of time with the tail between the legs and the waist closures done up. When the smock is worn open and the tail down it allows more freedom of movement and is much more comfortable. To his bottom right is the waist adjustment snap. To get a better idea of how the smocks are assembled, picture a large cookie cutter and each piece is cut out in mass. A stack of pockets, a stack of pocket flaps, a stack of fronts, backs, collars, etc. ... Pieces are simply assembled and sewn together according to pattern and not color arrangement. This explains the nicely colored pockets and the 'spilt paint' front.

Chapter Five: Issue Clothing

Left: The angola wool lined collar. Collarless shirts are also made of this material which can be seen underneath the smock. Note the variation in color of the angola wool. Right: One of the two shoulder straps. Each is secured by a plastic button as those also found on early Austerity battledress.

Left: The air vents found underneath the arms. There are three on the front half and four on the back. Right: The tapered sleeves end with knit cuffs, a tell tale sign of the 1st pattern smock.

77

For King and Country: British Airborne Uniforms, Insignia & Equipment in World War II

Left: A better look at the front tail snaps. The three rows are for the adjustment of the tail to fit the wearer comfortably. Right: Here the waist tab is undone. Note that when it is fastened it takes up about six inches of the waist.

Left: The half zipper of the first pattern. Most were made by Lighting Zip Company. Also of interest is the zipper's pull tab. Scraps of material are gathered and made into these pull tabs by folding them over, pulling them through and then sewing them in. Right: Here on the inside is the size label – size 8 was the largest made. A man larger than that did not jump as he would more than likely hurt himself upon landing. To the left of the label is the War Department stamp and Broad Arrow. Most, if not all, 1st patterns found are dated 1942.

Chapter Five: Issue Clothing

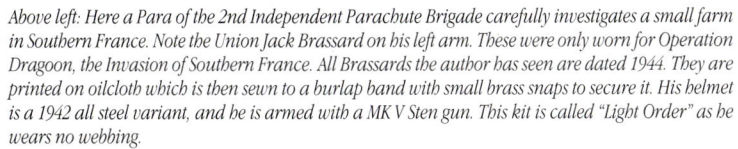

Above left: Here a Para of the 2nd Independent Parachute Brigade carefully investigates a small farm in Southern France. Note the Union Jack Brassard on his left arm. These were only worn for Operation Dragoon, the Invasion of Southern France. All Brassards the author has seen are dated 1944. They are printed on oilcloth which is then sewn to a burlap band with small brass snaps to secure it. His helmet is a 1942 all steel variant, and he is armed with a MK V Sten gun. This kit is called "Light Order" as he wears no webbing.

Above right: Another unusual 1st pattern smock which looks as is they wanted to conserve paint. This has been converted to full zip using as always the Jacket Parachutist's zipper. Note the smaller front snap panel compared to those shown previously. Tapered sleeves and knit cuffs ... 1st pattern!

Right: As mentioned, most if not all 1st patterns will be dated 1942. The red X is a bit of a mystery. Some say it was to signify that a piece had been issued. Others say that it is to signify that a piece was dropped from the Quartermaster's inventory.

79

For King and Country: British Airborne Uniforms, Insignia & Equipment in World War II

Above left: The lower waist tab is left undone. By the base color this one is perfect for North Africa, Sicily and Italy. The Rifle is a No.4 which was first issued for North Africa.

Above right: A back view. See how the color blends into the surroundings – perfect! Note the lack of rear tail snaps and how the tail simply dangles ... 1st pattern! Some 1st patterns have had their tails removed by the Paras as it became irritating for it simply hung there. So some daring individuals tore them off! Remember, this is after all King's property!

Right: A trooper gazes out and down unto the valley below. Of interest is the brush stroke pattern used here. This is a very unusual 1st pattern scheme as most examples show splotches and/or all colors running together, however this example is a series of curved 'snake' trails. A very special piece.

80

Chapter Five: Issue Clothing

Left: A Rifleman sneaks up to a machine gun bunker in hopes of flushing them out and avoiding a fight as he's out of grenades. Here you see the 3rd pattern 1943-45 web chin strap helmet and armed with the trusty No.4 Rifle. Right: A Brave Para tries his luck at bluffing an MG crew out of their bunker. Again in "light order" wearing the Toggle rope only. Note the interesting camouflage pattern that has been applied to this smock, and the blancoed anklets.

2nd Pattern

2nd Pattern smocks were made much differently than the 1st patterns. An artist made a drawing of the camouflage pattern and then a screen was made from this artwork. Larger screens were made compared with the German and U.S. camouflage patterns which repeat each 30". This was to better disperse the patterns. As you will see with the 2nd patterns most are of a standardized scheme compared with those of the 1st patterns which have very little standardization in their camouflage scheme.

Both the base color and layering camouflage colors of the 2nd pattern are very different from those found on the 1st's. The second pattern's base color varied from a sand to a light yellowish olive combination. The layering colors were reddish browns and bright pea greens to a dark olive. These colors were better suited for the Northwestern European theater.

The 2nd like the 1st was non-reversible and windproof though not waterproof. Second patterns usually had brass half zippers whereas the first pattern had the half steel zipper. This could simply be a manufacturer's preference. Air holes were sewn underneath each arm pit for ventilation. (there are usually three on each side). An "Ape Tail" was sewn to the back which could also be worn between the legs like the 1st pattern yet, the 2nd pattern had two snaps on the lower back allowing the tail to be secured and worn up. This is one of the big differences between the two patterns.

Second Patterns were of a similar cut to the first, featuring a half zipper plauquete front like the 1st pattern. These steel zippers were made by the Lighting Zip Company. Also featured were two breast, two waist, and two internal pockets, and two

A 2nd pattern Denison is shown here. Note the wrist tab adjusters and the non-tapered sleeves. Also a dead give away is the lack of wool knit cuffs. Also note the major color variation from the 1st patterns. The colors here are best suited for NW Europe as the 1st patterns seem to blend in better with sandy terrain.

shoulder straps. Yet the sleeves of the 2nd pattern were of a looser fit and not tapered to the wrist like the 1st. The 2nd patterns also had in place of the wool cuffs a wrist tab and three buttons which allowed the cuff area to be tightened or worn loosely. Those 2nd patterns found with hose tops added are post-war additions. Remember that these smocks were worn well into the post-war years.

Left: Note the very reddish brown and pea greens found on this piece, and that the zipper pull tab is missing.

Chapter Five: Issue Clothing

Right: Here can be seen the second difference between 1st and 2nd patterns, the presence of rear tail snaps. The tail can be folded up and secured to the back. Note that one brass snap is missing from the tail.

Below: A better look at the right side's unusual brush strokes compared to those on the left giving it a very uneven look. As previously mentioned they are sewn together according to the pattern and therefore are not fashionably correct.

Below right: A look at the two sides together. Very uneven, yet this also helps to break up shapes and conceal the form of the wearer.

83

For King and Country: British Airborne Uniforms, Insignia & Equipment in World War II

Left: Note the 2nd pattern's wrist tab. The tab can be buttoned at one of three positions – the first when the most room is desired. Right: The second button tightens the wrist for a closer fit.

Left: The third button is the tightest and smallest that the wrist can be set. Note the loose fit as is. Right: And here the underside of the tab. Note the many colors especially those where one brush stroke goes over another making yet another color.

Chapter Five: Issue Clothing

Denison Variants

There are three known variants that have been modified from the manufactured, issued smock. Note that many additions are possible. Of interest is that a unit rigger or tailor could make special custom modifications. These are very much an exception, and photographic evidence is limited.

Experimental

There was an experimental 2nd pattern smock that was labeled: Smock Airborne Troops Experimental. The difference being that in place of the brass half zipper there were six brass buttons. These did not see wide use in the field as the buttons would pull off and were not as sturdy as the zipper. Thus it was discontinued. Note that plastic buttons were also found on these trial variants.

Snipers Smock

Snipers are well known for making special modifications to both their uniforms and equipment. The Canadian Military Headquarters actually made one such modification official, this was the adding of a rear pocket to an issue Denison. This modification was at the CO's discretion. This pocket could be used to carry either a water bottle, as when sniping, the rifleman would not wear his webbing, binoculars if not worn around the neck, or ammunition. Of interest is that the 2nd Battalion of the S. Staffs wore this variant of smock in Arnhem.

Full Zip Smock

Issued as a half zip and then converted to full zip. The conversion was done by removing the half zip and replacing it with the long full zipper from the olive denim Jackets Parachutists. A full zip was much easier to get in and out of than the half zip. These did not see wide use during the war. The author has seen photographs of both generals Gale and Browning wearing full-zips, these two being an exception and considerd an officer's modification. I am of the opinion that these smocks were converted in mass after the war, and then issued in mass. Photographic evidence does not support the use of the full zip during the war, besides the very limited exceptions as mentioned.

Officer's Smock

Officers were known to have specially custom lined collars added to their Denisons for added comfort. Usually they were lined with a corduroy material. Officers also were known to have full length zippers added to their smocks (these were limited to general officers from what I've seen).

Jackets Parachutists 1942 Pattern

The Jackets Parachutists first saw use in 1942 and was designed to be worn over the Denison and all of the web equipment with the parachute harness worn over the Jackets Parachutists. Its purpose was to keep all the equipments in place while in mid-air and to reduce air resistance. Made of olive denim and featuring two grenade pockets, an "Ape Tail" with three sets of 'Newey' brass snaps for adjusted comfort, and a full length zipper. This very zipper is what was cannibalized and used to convert the Denisons to full zip. Upon landing, this oversmock was discarded. This sleeveless garment came in three sizes, one small, two medium and three large. This was issued and worn by most Paratroops, though some did jump without it as they thought they would be able to get into action sooner if need be without the over garment to deal with.

Shown here is a special Canadian variant. Note the added bomb pockets. These have been removed from the Jacket Parachutist which was somewhat common among the Canadians. Note the Canadian jump brevet on the left breast. Also note how far forward the waist snaps have to reach to tighten the smock.

Right: The size label of the Jackets Parachutist commonly known as the Over Smock. The author is aware of 4 sizes, and at 5'7" jumped wearing this one and it dwarfed me even with webbing on and chute and harness. Of interest is the very faded WD stamp just below the label.

For King and Country: British Airborne Uniforms, Insignia & Equipment in World War II

The Jacket Parachutist. Note its large oversized fit – this allowed for it to be worn over the Denison and also over the P-37 webbing. Of interest is how the MK V Sten is broken down and carried under the harness.

Shown here is the zipper and pull tab of the Jacket Parachutist.

Here can be seen the Jacket Parachutist and Denison side by side. Note the variation in wool lined collars.

86

Chapter Five: Issue Clothing

Footwear

Ammo Boots

Ammunition Boots were made of pebbled leather and dyed black. They also featured a toe cap for further protection. They also featured a leather sole with heel cleats, toe plates, and hobb nails. Date of manufacture and the manufacturer can be found stamped into the inside of the leather insole as well a the broad arrow and WD stamp. Some WD and broad arrows are stamped into the top ankle ridge. There were three fittings to each size: 'S' for small, 'M' for medium, and 'L' for large.

Hobb nails were hammered into the soles and done so in four patterns. The first was five rows of five nails. On April 18th 1942 the nails were reduced to three rows of five. On August 29th 1942 the number were again reduced to thirteen nails. This being from the toe, two, three and then two rows of four. All in aid of the war effort as these nails were large, about 1/2" round. These were used up until 1944 when smaller nail studs were used. There was also a shortage of screws for the toe plates. So on September 2nd 1942 one screw was deleted from further productions. This did not hamper the boot nor loosen the plate. Many a collector has, I am sure, wondered why one of his screws were missing, and now they know. Leather laces were worn with the boots. By regulation they were 'ladder laced'. And only in this manner were they worn under strict regulation.

Officers were allowed the privilege of wearing brown boots. These were by private purchase which would be obtained from either a regimental tailor or an approved outside tailor.

Note how the laces are done. This is the ladder lace by regulation method – this was how the laces were to be tied, and it was followed! You can see the toe cap and how the P-37 anklet is designed to fit on top of the boot and helped blouse the trousers.

Note how the trousers blouse in the web anklets and the pebbled leather of the boot. Again this pair is ladder laced.

The hob nails which were worn throughout the war, from the 1930s until 1945 when replaced by the smaller nail studs. This pattern seen is the last pattern worn before the late war nails came into service. This pattern was approved on August 29, 1942. Also of interest is the missing toe plate screw on each boot which was an effort to save materials, and was approved of on September 2, 1942. Note the stamping on the arch, broad arrow, size 13 small.

Right: Note the WD broad arrow stamping on each boot.

For King and Country: British Airborne Uniforms, Insignia & Equipment in World War II

Left: On most war time boots the inner sole is stamped as shown here: manufacturer, date, size (13 Long). Right: Tea Time! Note the heel iron and toe cleat – each pair had them, as well as hob nails. Shown here is the late war 1945 pattern, as the earlier war studs were much larger by about 1/4" round – this pair is stamped on the arch. The manufacturer and Broad Arrow is just visible. Note the tea cup is post-war as this color was not issued until the 1950s, yet it is an interesting shot of the boot.

Socks (Hose)

As with many of the undergarments, there are several variations of woolen socks. The most common are the slate gray socks which are full length and came up just below the knee. On one sock, of each pair, was a size, date and manufacturers label which was sewn to the sock. These socks also came in a light olive color. A lucky soldier would have two pair, the extra being carried in the haversack. For darning and repairs there were two small balls of wool issued and kept in the soldiers housewife (sewing kit), this was carried in the haversack. The original issue was three pair yet on November 1st 1941 this was reduced to two pair.

Hose Tops

Think of socks that have had the foot cut off, these are hose tops. These were part of the tropical issue kit and were widely used in North Africa, Sicily, and Italy. The idea and practice was for the wearer to slide the hose top over the sock, wearing it over the ankle and shin, this helped to keep sand out of the soldiers boot. As members of the Airlanding Brigade wore KD (Khaki Drill) shorts in Sicily for example.

Puttees

Puttees were long strips of serge wool with long cotton tapes attached. They were wrapped around the ankle and shin. They were further insurance to keep sand out of the soldier's boot.

Each man was issued with two pair – one to wear, and a clean spare to carry in his haversack. Each pair had a size label sewn to it as shown here. Socks varied in color, some were white, some khaki – most common were the gray. These were full length and went to the knee.

Right: The size label. Like most labels it featured date, size and manufacturer. The author is aware of three sizes, possibly four.

Chapter VI

Headgear

Sorbo Training Helmet

The Sorbo was the first pattern training helmet worn by the fledgling Paratroops. Of a simple design and made of formed slabs of Sorbo rubber glued together, it was able to fill a gap for head protection until the better constructed Bungy helmet was designed and then put into service.

Bungey Training Helmet

The Rubber Bungey was designed for training purposes. It was light weight and constructed of canvas which covered a foam rubber crown. Also featured was a neck flap that covered the ears and tied together underneath the chin. It was introduced in 1942 and used throughout the war. There are three variants that the I have seen, one with metal grommets placed along the ear section of the chin straps and a second without the grommets. A third version had slit holes without grommets. Various weights of twill were used in the construction of these helmets – these were simply the choices of the manufacturer and the materials that were available.

Shown here is a Recce trooper with his fellow crewmen in a roadside ditch. Note the chamois lined chin cup of the 1943 all steel helmet. At the rear is a driver wearing the namesake of the Airborne – the maroon beret.

This is the very rare first pattern Training Helmet. This helmet was an inexpensive stopgap variant which was replaced by the proper canvas covered Bungey Helmet. Also of interest is the very rare converted for Airborne use MK II Gas Respirator. It is shown here upside down so the trooper could simply and quickly unsnap the bag and the mask would fall out ready to use in case of gas attack. This Para is wearing a 'Step In' gabardine smock. Note the jump wing, and the Worsted Corporal chevrons worn on the sleeve.

Taken at Ringway in May 1942. Here is a group of trainees, all of which wear the Bungey training helmet and 'Step-in' gabardine smock. All but one wear a P-37 waist belt as per regulation. On the far right is Colonel (then lieutenant) Sweeney of the 2nd Battalion, Oxf & Bucks. On each man is the Quick Release box harness with 'X' type parachute. Also of interest is the stenciling on some of the Bungey helmets, P.T.S. (Parachute Training School).

A Trooper holds the 2nd pattern training helmet known as the Bungy Helmet. It was made of a thick foam crown covered in cotton twill with a tie flap on each side, and was a simply manufactured piece of equipment. Note the size label inside – though there may be more, the author is only aware of sizes 1 and 2. There are possibly four variations of this helmet. This Helmet has no slits or grommets in the tie flaps. Please note that terms such as "tie flap" are designations by the author as there are no known official descriptions for such pieces. Note that the flap ties are secured by simply tying them in a knot as there is no buckle. All Bungey helmets are as seen here, without a buckled chin strap.

A look at the inside of the Bungey helmet. Note that the top inside lining has a sewn in leather adjustable tie – this will enlarge or decrease the size and fit of the helmet. These helmets appear from time to time in the collector world, so note that they are easily copied and sold as authentic.

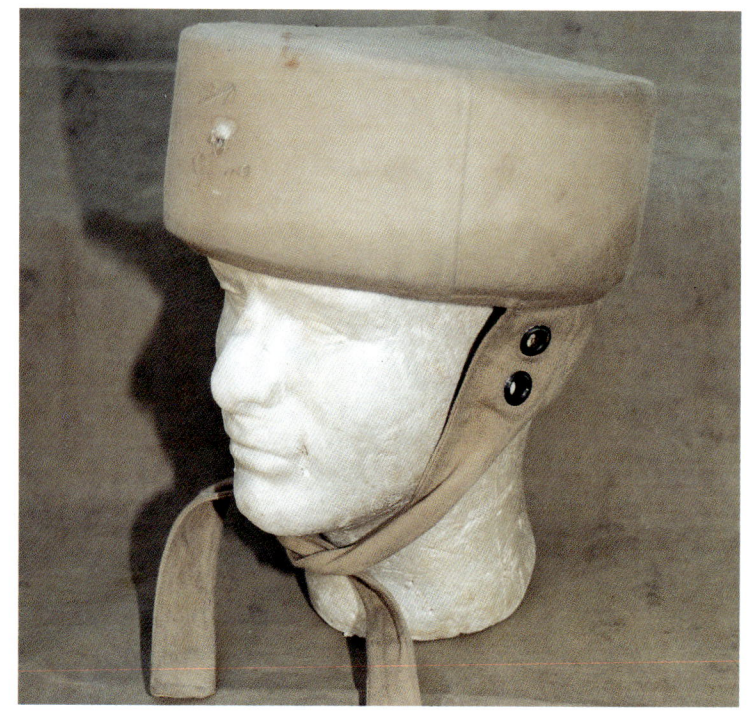

Right: A Bungey variant with metal grommets. As the author has seen others with larger brass grommets, these must be a difference in manufacturing design. On the front crown of the helmet can be seen where a badge was once attached – one may think Polish, yet the Poles stenciled the eagles on their training helmets. When worn for a jump the chin strap would be knotted again and then wrapped around itself – if kept as is the helmet would go flying off as soon as a para jumped out of the balloon or aircraft.

Chapter Six: Headgear

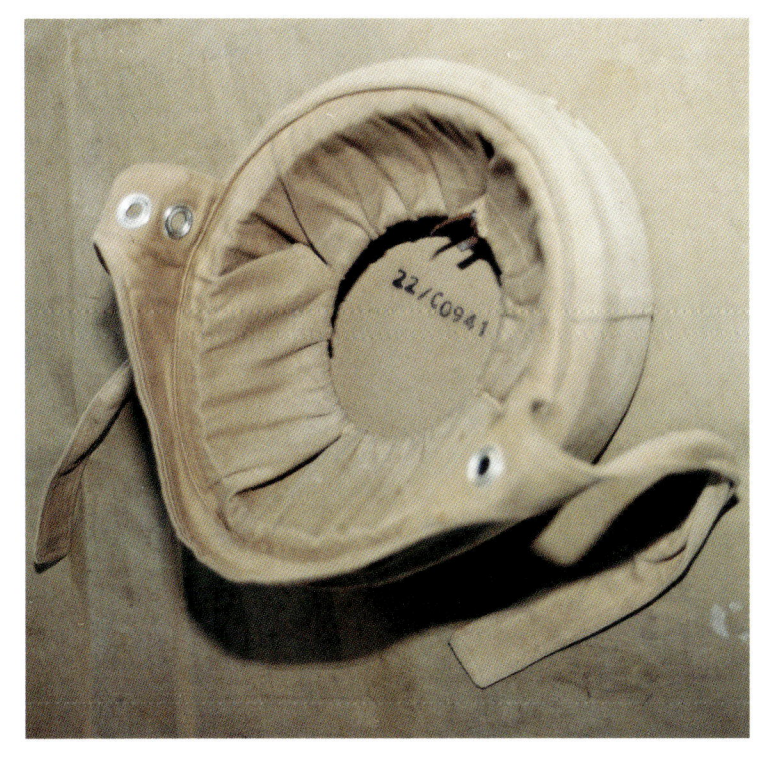

Above left: The interior – note that the grommets are not painted on the inside. The number stenciled could be a man's army number. Note the leather tie used to draw up the helmet's inner liner.

Above right: The piece of rubber that makes up the helmet's inner crown.

Right: A better look at the grommets. These are to improve one's hearing, rather than hearing a muffled rumble.

Side by side are the late war 1943-45 web chin strap helmet and the fibre rim 1942 helmet. Again it was all about saving material and money badly needed for the war effort.

Jump helmet, Prototype

The 1941 Prototype helmet featured a hard rubber rim fitted to the edge of the manganese steel shell which came to a lip that protruded from the back of the helmet. This helmet saw little use in combat as it was soon replaced by the fibre rim helmet, though it was worn for the Bruneval Raid in February of 1942. Of interest is that two 1st pattern Prototype helmets have been found buried in Holland, near the Hartenstein Hotel which is now the Airborne Museum. This goes to prove that exceptions to the rule do exist, but do note that this for instance is most unusual.

Jump Helmet, 1st pattern, Fibre Rim

Featuring a flat 1" hard rubber rim running along the rim of this steel shell and was worn by the British, Canadians and Polish troops, and was first issued (and manufactured) in 1942. It featured a black leather chin strap which was attached to the shell via four screws, two in the rear, and one on each side. The chin strap cup was chamois lined. This helmet was used throughout the war and can be seen in use in Arnhem, though in small numbers as by then it was old stock and manufacture had ceased in 1943.

For King and Country: British Airborne Uniforms, Insignia & Equipment in World War II

Above: This is the 1942 Fibre Rim Helmet (made only for that year yet worn throughout the war), which was commonly worn for the Arnhem Operation by the British and the Poles. Note the screw which holds the band, liner, sweatband and chin strap in place. There are four such screws – one on each side, and two side by side 1" apart on the middle back.

Right: Interior of the Fibre Rim helmet. Note the foam used as cushioning (which by the way works very well – I know, I hit my head on my second and last jump. This helmet saved my life – Author). Note the steel band between the shell and foam which keeps the helmet together and acts as a type of gasket. The string and cotton twill suspensions adjust the position of the helmet on the wearer's head. Also note how the rear chin straps cross over to attach to the helmet. This is to better secure the chin strap and fit on the wearer as these are worn tight when jumping.

Right: The manufacturer stamp, size and date stamp. Some common helmets have the last number of the date stamp rubbed off. This is very likely because it is post-war, as the web chin strap helmets were made well into the late 1940s. Reproduction war-dated liners are abundant, and post-war liners are also placed into early shells. As a note, each shell is usually stamped as well.

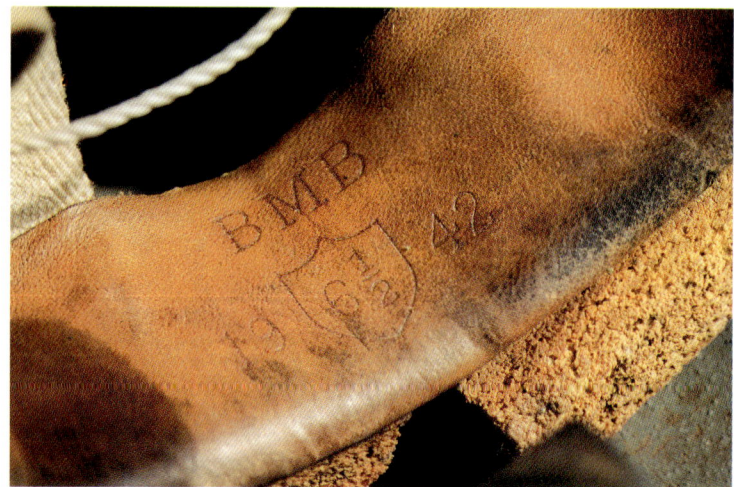

Jump Helmet, 2nd pattern, All Steel

Featuring a manganese steel shell with the same black leather chin strap as the fibre rim. Manufactured and issued in 1942/43 this helmet was worn mainly by the British in Italy, Normandy, Arnhem and Germany. The 2nd pattern is the most common of the leather chin strap helmets.

Left: The All Steel Helmet. Note that on this helmet the leather chin strap is riveted. The later war 1/4" net is worn over the helmet shell with scrim adorning in the best traditions. A .303 bandoleer is slung around the shoulder as the trooper is in light order and is armed with the No.4 Rifle.

Chapter Six: Headgear

Right: The inside of the All Steel Helmet. There is no difference between this and the Fibre Rim, except for the fibre rim. Again, notice the chamois lined chin cup. Also note the grommet holes at the back of the helmet on the chin strap. These are to raise or lower the chin strap via the screws for a proper fit. All leather chin straps had these grommets. There are many copies on the market today – and this is one of the things they often forget to add. One last note is how the helmet net is secured by the pulling of the net string and tying in a knot. Some nets were larger than others and had to be tucked in between the shell and foam.

Below: The manufacture size and date stamp. These All Steel were made as early as 1942. Most stamps seen by the author separate the year by placing the size between it. Exceptions known to the author are the C.W.L. and C.C.L. manufactured web chin strap helmets, dated 1944.

Jump helmet, 3rd pattern, Web Chin Strap

First issued and manufactured in 1943 and used well after the war. This pattern featured a manganese steel shell with an economy web chin strap as leather was expensive and in short supply. These first saw action on June 6, 1944 for the Normandy Invasion. The fourth pattern's web chin strap was secured by three screws, one on each side and one in back to secure the liner and then held in place by three rings attached to the liner. The web chin straps were difficult to put on and take off, though the Government wasn't too bothered about the soldiers comfort (rather saving money). This helmet was issued mainly to the Canadians, Poles and the French. Two out of ten British Paras wore the this helmet in combat.

Left: And lastly is the third pattern Jump Helmet (third pattern as it was the third pattern helmet to be issued and put into wide use), a late war 1943-45 web chin strap helmet. This is the most commonly found jump helmet by collectors. Note how the chin strap secures and buckles to the left side. The chin strap can be unbuckled as it is suspended by two buckles, one on each side and one loop in the rear. The chin cup is chamois lined and the front is leather reinforced. On the late war helmet there are three suspension screws – these hold the two buckles and one loop in place as well as the steel band and liner. Each helmet has tiny bits of cork added during the painting process to help break up the otherwise shinny surface. Right: Two late war helmets. To the left is type issued to and worn by the Polish as can be seen by the golden eagle painted on it.

For King and Country: British Airborne Uniforms, Insignia & Equipment in World War II

Left: Here a Para tightens his chin strap. Note how the paint wears off the rim from wear. Right: Note how the chin strap is positioned on the face and jowl. This was by far more difficult to put on and take off compared to the leather chin strap helmets. Note also the Battledress Serge 1940 Pattern blouse.

Here is the liner and chin strap assembly out of the shell. Note the liner is made up of a rim, a thick piece of foam, and leather head piece, which are all held in place by a piece of leather cording woven between the band and the cotton tapes of the leather liner.

Here is one of the two buckles found on the late war helmet. Also note the top inside pad which added protection for the top of the head.

94

Chapter Six: Headgear

A better view of how the band is attached to the foam and leather liner via the leather cording which runs through the middle on the band. The screw hole is one of three which hold the liner into the shell.

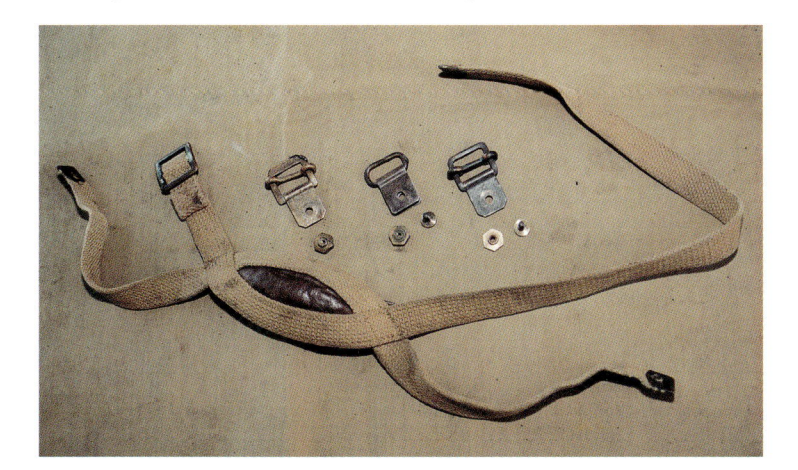

The chin strap assembly. Note that the screw of the far left buckle is missing. On each side are the buckles which are found on each side of the helmet, and in the middle is the buckle that is found to the rear of the helmet. It is different as the rear buckle does not adjust, but simply holds the web strap in place.

Shown here is the commonly found stamping on the leather sweatband, and is stamped as follows: manufacturer, size, and date. Note how the date stamp is rubbed out. This is often done to post-war helmets so they might pass as war time. Remember these helmets were made well into the post-war years.

The leather inner crown pad. Just visible is a white broad arrow stamped onto the pad – this is unusual as most, if not all, stamping is made into the leather sweatband.

Helmet nets

Two variants were predominately used, the 1" and 1/4" square nets. Each had its own tie string which secured the net to the helmet. Some Paras placed burlap sacking underneath the netting for added camouflage. Scrim would then be tied and woven under and over the individual netting squares. Scrim consisted of pieces of colored burlap which were cut into various sizes, and then applied to the netting. Colors came in chocolate, dark olive, light olive, red brown, and khaki.

Looking much like a combative bush this Para has decorated his helmet in true Airborne tradition. Each man was issued a net for his helmet. This was to also help in breaking down the shape of the helmet and deflecting any glare from the sun.

95

Flying Helmet

A special flying helmet was designed for the Glider Pilots to wear in flight. It consisted of a protective shell with earphones and a mouthpiece which attached to each side of the helmet. The mouthpiece enabled the glider pilot to communicate with the transport aircraft. Upon landing they would don an issue jump helmet for combat duties. Remember that this was a part of their role, that they were to serve as infantrymen once they completed their pilot's tasks. Each Glider Pilot was issued with two helmets, one flying and one combat (jump). Leather Flying Helmets were also worn, more so in training.

Beret, Maroon

"There's nothin' like walking down the street wearin' my berry, it's a feelin' of pride! There's nothin' like it my boy! (veteran Wally Walsh [my Uncle] 1st Parachute Bn.)." Affectionately called by those who wear it, the "berry", the Maroon beret is the signature of the British Airborne Forces. In August 1942, prior to the formation of the Parachute regiment, Major "Boy" Browning wanted a distinctive piece of headwear for the Airborne Forces. Two colors were offered, light blue and maroon. The beret first saw service in North Africa, November 1942.

Veterans were known to customize their berets by shaving the wool for a smarter look, gutting the liners for a better fit and either shoving in the string tie (once the perfect fit was achieved) and or tying a bow and then sewing it to the leather sweatband. The string tie did not hang down, this was frowned upon as it was a 'French Thing'. War production berets are 1" to 1 1/2" larger than post-war manufacture.

How the net is secured. The green string fed through the net, pulled taught and then tied. Also noteworthy is that this chin strap is sewn and not riveted. Note where the brass loops are held in place and where the 'L' of the strap meets and is sewn together. As mentioned previously some were also both sewn and riveted. This helmet is a B.M.B. 1943.

Here it is the Namesake of the British Airborne, the maroon beret. Note the brass air vents to the right and the tie string in the rear. This string was tied in the desired position and then either pushed into the leather sweat band or cut off or sewn to the sweatband – it did not simply hang down as this was considered a very "French thing" to do and was frowned upon. Of interest is the white slotted diamond stamp which is found on most examples. Inside the white slotted diamond stamp are the markings: size, manufacturer, broad arrow and year of manufacture. Kangoal Wear Limited were the predominate manufacturer of berets, and not only for the Airborne but also for the Royal Tank Regiment (black), and the Commando (green). The paper tag at the rear of the beret was an issue tag and its details were written in pencil – few of these paper tags remain today due to wear.

A Recce driver taking cover in a ditch. Note his plastic Recce badge on his beret. Berets were commonly worn in combat among the Airborne.

Chapter VII

Airborne Insignia

Cap Badges

There were four types of badges manufactured and worn during the war. For other ranks there was brass, white metal, and plastic economy. Officers also wore white metal and bronze (from their parent regiment) and were allowed to wear private purchase badges in sterling silver. Plastic badges were in aid of the war effort – for example saving brass for the manufacture of shells. Plastic badges are very rare as they are not as hard wearing as metal. Also, the troops would throw them away and replace them with metal badges as plastic did not look "smart." The British Army takes great pride in turning out (to come out in immaculate dress) smartly. Plastic badges cannot be polished like the brass and white metal. Therefore the plastics did not fare well and are now very collectible. All badges had what is known as a "Kings Crown" nicknamed the 'dog and basket' (as found atop the Parachute Regt. badge). The Queens Crown is unlike the Kings crown and is inverted in the middle like a pair of breasts. This Queens crown was first put into service in 1953 and is still in use and will remain so until a new sovereign adopts a new crown.

Here are some of the goodies that we badge collectors go mad for.

For King and Country: British Airborne Uniforms, Insignia & Equipment in World War II

Parachute Regiment

There were three variants made during the war for the Parachute Regiment. The first was the plastic economy cap badge made of a silver colored plastic and attached to the beret by twin brass keepers and pin. The plastic badge first saw service in early 1943 replacing the Army Air Corps badge as the Paras were part of the Army Air corps before becoming their own regiment. The plastic badge was first introduced on August 20, 1943 and was discontinued on May 9, 1945. Some 118,456 plastic badges were manufactured.

A second badge was made and issued in white metal for both officers and other ranks. Sterling silver badges could be purchased by officers. This badge attached to the beret via a brass pin and lugs. Officer's collar dogs of the same design were made to be worn on their Service Dress. This was a gabardine wool uniform made by either a Regimental or local tailor and not worn into combat.

A third variant of the cap badge first saw service in 1944 and was worn into the post war years until the Queens Crown replaced it in 1953. This late war badge was made of white metal and featured a slider in place of the pin and lugs. This slider slid through a small slit cut into the beret.

Of interest are the following Battalion practices: the 5th Scottish Parachute Battalion (2nd Parachute Brigade) wore Balmoral bonnets with the A.A.C. badge 'backed' with their clan tartan which was Hunting Stewart; the 7th Parachute battalion backed their Parachute Regiment cap badges with a diamond of rifle green felt as they were originally Light Infantry prior to becoming Airborne.

Badges of the Parachute Regiment. From top to bottom and left to right: the reverse of the late war pattern badge as it has a brass slider. The front of the late war badge. In the middle is the Plastic economy badge. Below the plastic badge is the mid war badge with the pin and eyelets. Pins and eyelets are the mark of a good World War II badge as the sliders came out later and were issued into the post-war years until it was replaced by the Queen's Crown badge. On the bottom row is the Officer's Collar Dogs which are worn on the Service Dress uniform.

Chapter Seven: Airborne Insignia

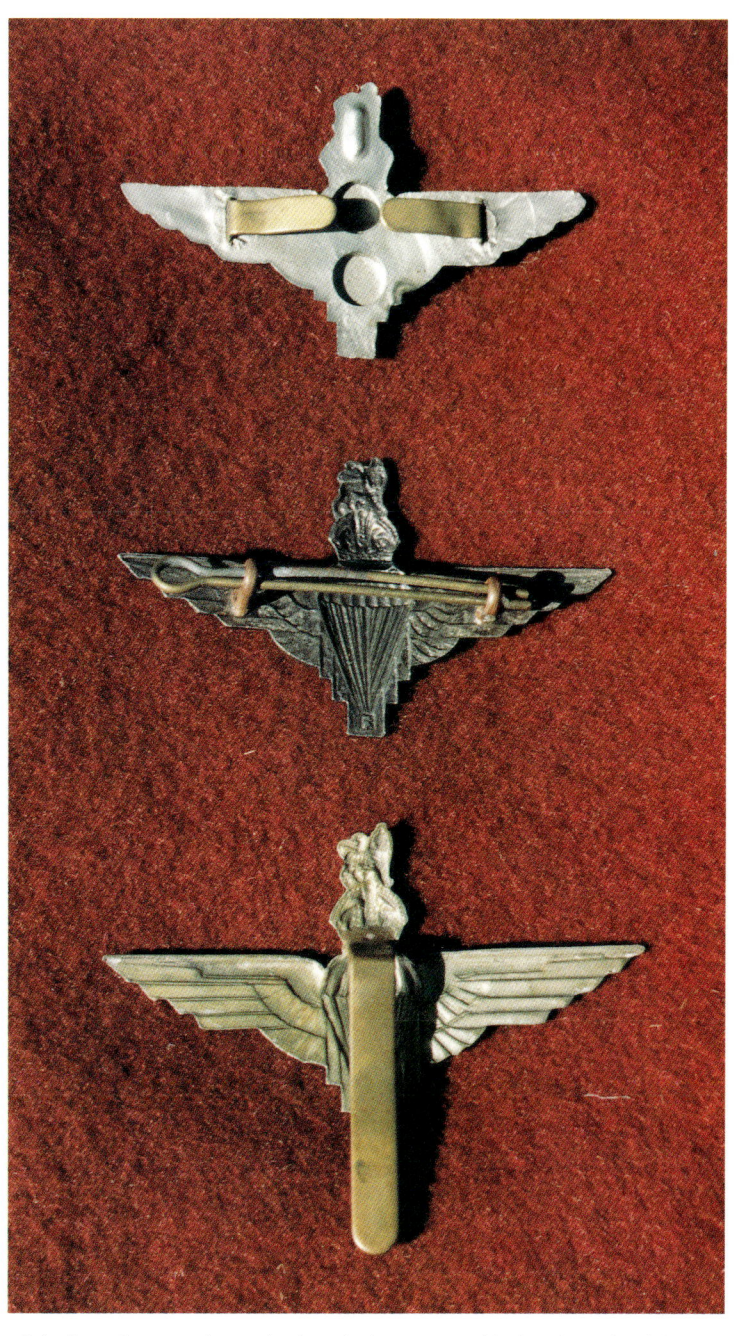

Left: Top to bottom; Plastic, an Officer's pin and eyelet badge and lastly the slider. Note that with the exception of "the plastic" these are nicknames that the author has given to each badge to better identify them and are not the official terminology. Note the detail of the Officer's badge. 'Re-Strikes' are post-war badges that have been cast in the same dyes as those used in the war. In fact they are the same dyes, yet the badges are not of the same materials and are of lesser quality. Be aware as restrikes can easily be mistaken for the real thing as it is very difficult to tell the difference. One determining factor is that the restrikes weigh less than original badges. Right: Top to bottom; Plastic badge with its brass keepers. These do break off with wear. As can be seen it is a simple molded piece. These brass keepers are found on 99% of the plastic badges. This badge was introduced on August 20, 1943 with production ending May 9, 1945. Some 118,456 plastic Parachute Regiment badges were manufactured. In the middle is the pin and eyelets badge of 1943 (which better secured the badge than the sliders). On the bottom is the slider badge of 1944. There was a special brass clip that fit over the slider to keep in from sliding out few of these are found today.

Air Landing Brigade

All members of the Air landing Brigade wore their regimental cap badges on their maroon berets. The Oxf & Bucks, being of a Light Infantry origin, 'backed' their badges by placing a small piece of rifle green colored felt material behind the badges. The RUR, being a Rifle Regiment, also backed their badges with Rifle green felt which is one of the colors of the Rifle Regiment (black and green or black and red). Plastic badges were not widely worn by the Airlanding units. Plastic badges I have seen are the Border, K.O.S.B., RUR, and Devons.

The Kings Own Scottish Borderers shoulder title which was worn by the 7th battalion, yet in limited number. Beneath the title is the embroidered Pegasus flash and Airborne strip, white metal cap badge and glider badge. Predominately, the 7th Battalion K.O.S.B. wore a flash of Leslie Tartan upon their BD in place of the shoulder title. Also in evidence is a complete lack of title or flash, simply leaving a large space at the top of the sleeve, then beginning with the Pegasus and AIRBORNE strip. Also note that the AB strip is a hand sewn copy as an original was not available. There are plastic K.O.S.B. badges, yet they are very rare.

Here is the worsted shoulder title, printed economy Pegasus flash and Airborne strip. Below this are the Devon badges, on the left the bimetal (brass and white metal) and right the plastic. It is possible that the 12th Battalion Devons also wore the printed DEVON title.

Shown here are the 2nd South Staffords shoulder title, embroidered Pegasus flash and Airborne strip, bimetal cap badge and glider badge. The title featured with golden seraph lettering on cherry red back is an unofficial title and worn only by the 2nd S.Staffords Air Landing Battalion. As of this writing, the author not seen a plastic S.Staffs. cap badge. Note that this unofficial title is a copy.

The Border embroidered shoulder title, embroidered Pegasus and Airborne strip, plastic Border badge (that the author was told didn't exist), white metal badge and the glider badge. To the author's knowledge the 1st Battalion Borderers did not wear a printed shoulder title.

Chapter Seven: Airborne Insignia

The Royal Ulster Rifles embroidered title, embroidered Pegasus and Airborne strip, white metal badge and glider badge. There also exists a very rare plastic cap badge. To the author's knowledge the 1st RUR did not wear a printed shoulder title on their BD. This, like the S.Staffords, is an unofficial unauthorized title yet it was specially made for and only worn by this Air Landing battalion.

The 2nd battalion Oxf & Bucks embroidered shoulder title, worsted Pegasus and Airborne strip, white metal badge and glider badge. To my knowledge the 2nd Battalion did not wear plastic cap badges or printed shoulder titles. This title is unofficial and unauthorized yet specially made for this Air Landing Battalion.

Colonel Sweeney in May, 1942 with the rank of Subaltern (2nd Lieutenant). The FIFTY SECOND shoulder title is shown here worn in conjunction with the Pegasus and AIRBORNE strip. Note the gorget worn on his collar, this was done by Colonels and the General staff, yet the officers of the 2nd Battalion Oxf & Bucks wore this 'self made' gorget on their collars, from the rank of Subaltern and up. These self made gorgets are made of the smaller Oxf & Bucks Regimental buttons like those found on the Field Service and Service Dress caps. These were sewn onto a 3mm brown colored cord that was looped around the button and then went down and under the collar.

For King and Country: British Airborne Uniforms, Insignia & Equipment in World War II

Corps and Administrative troops
All wore the maroon beret with their own individual badges. As with the Parachute Regiment and Air Landing Brigades there were brass, white metal and plastic. All saw use throughout the war.

Right: The embroidered Recce title, printed Pegasus and Airborne strip, brass and plastic cap badges – the brass badge was for Other Ranks, and bimetal for Officers. Note that both printed and embroidered titles were made for all the Divisional troops and worn with both the printed and embroidered Pegasus and Airborne strips.

Embroidered Corps of Military Police title, printed Pegasus and airborne strip, brass and plastic cap badges. Note that the R.C.M.P. title is post-war as they were not designated Royal until after the war.

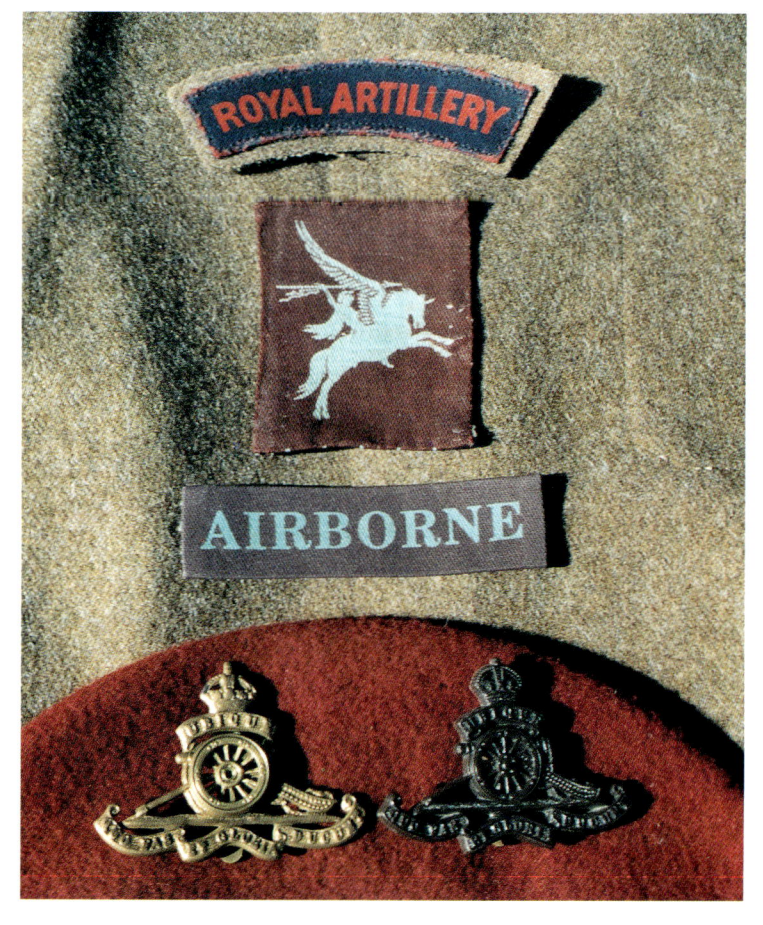

Right: The printed RA title, printed Pegasus and Airborne strip, brass and plastic cap badge. This is one of the rarest of the plastic badges as very few were manufactured. Note how the title is sewn to a backing of serge.

102

Chapter Seven: Airborne Insignia

Left: A printed Royal Army Service Corps title, printed Pegasus and airborne strip, brass and plastic cap badges.

Printed Royal Electrical and Mechanical Engineers title, printed Pegasus and Airborne strip, and the brass and plastic cap badges.

Left: Royal Army Ordnance Corps printed title backed on a piece of serge, printed Pegasus and Airborne strip, and the brass and plastic cap badges.

103

Left: Printed Royal Army Medical Corps title, printed Pegasus and Airborne strip, and brass and plastic cap badges. Of interest is the variant title which was spelt out Royal Army Medical Corps in yellow on maroon, was made only for the Airborne and is a very rare title. Right: Royal Armoured Corps shoulder title, printed Pegasus and Airborne strip, and the metal and plastic cap badges.

Left: Royal Corps of Signals printed shoulder title with the printed Pegasus and airborne strip, and the bimetal and plastic cap badges. This title is unofficial, as the ROYAL SIGNALS was the authorized title for wear. Right: Royal Engineers printed shoulder title, printed Pegasus and Airborne strip, and brass and plastic cap badges.

Chapter Seven: Airborne Insignia

Glider Pilot Regiment

Glider pilots wore only the A.A.C. badge during the war and were made of plastic, white metal (1943), sterling and gilt for officers by private purchase. The A.A.C. plastic badge was first introduced on May 26, 1942 and was discontinued on June 19, 1945. Some 77,928 plastic badges were manufactured. The Glider Pilot regiment wore the A.A.C. badge until 1958 when the Glider Pilot Regiment cap badge was introduced.

War Correspondent

This was a gold 'C' within a gold circle on dark green. This was worn on the maroon beret. I only know of this badge being made of worsted. Little is known of its origins.

Wings and Qualification badges

Jump wings

Jump wings were awarded to those who had completed two jumps from a balloon and five jumps from an aircraft. Those authorized to wear the wings were members of the parachute regiment and those corps and administrative troops who went into battle by parachute, these being the Corps of Military Police, Recce, R.S., R.E.M.E., R.A.M.C., and on. All had parachute trained and qualified troops who wore the wings. The wings were first introduced on December 28, 1940.

The wings were worn both on the right sleeve of the Smock and on battledress blouse. No insignia besides rank or wings (jump and glider) were worn on the Denison. Two exceptions are the Air Dispatch. and the Army Film and Photographic who both wore special flashes on the sleeves of their Denisons.

On the left is the white metal Army Air Corps badge and right the plastic. There were two variants of the white metal: one with slider as shown, and another with pin and eyelets. The plastic badge is held in place by a pair of brass keepers. The plastic badge was introduced on May 26, 1942 with production ending June 19, 1945. 77,928 plastic badges were manufactured.

The Author: "I had a wonderful visit with an ex-Glider Pilot a few years back. I say most affectionately that Harry is the epitomy of an English gentleman, especially in this day and age where manners and tradition are practices of the past. As we were leaving that evening his wife asked if we'd like to see his uniform, and of course you know we did not protest. And out she brought this grouping in a plastic shopping bag. Shown here is his Battledress Serge 1940 Pattern blouse and trousers. He was issued the Bush Hat as he was sent to Burma to prepare for the Invasion of Malaya. Note the printed Pegasus flash on his hat still as it was during the war. Harry said that it was so hot that when they were flying the Horsas the glue melted and the sides started to come undone. They had just touched down before the glider came completely apart. At the base they then waited for American made Waco gliders to replace the Horsas as they were now unsuitable and unsafe. Harry flew into Normandy, was wounded at Arnhem and went across the Rhine in this BD and beret. Note the plastic A.A.C. cap badge. The white diamond stamp is well worn out from wear. Harry was a 1st Pilot and held the rank of Staff Sergeant. All the shoulder insignia is printed, and on the left breast is the 1st Pilot's brevet. His ribbons were added after the war as they were not issued or worn during. What a bit of history you see there upon the table. We had a great time hearing of his experiences yet Harry requested that I not print any of his experiences and I must keep that promise."

Three variants of the jump wing. From the top: a very early variant; note the twin layers of feathers and the single color of white on serge. Also note the large half oval serge backing. In the middle is a fairly uncommon variant with twin layered feathers upon its wings. Note its half moon serge backing. Both the top and middle wings feature "lined and couched" parachute shroud lines. On the bottom is the most common wing seen worn by the Paras. Note that there was a wing on a black background which was only worn by the 156th Parachute Battalion.

There are several variants of the British jump wing that were manufactured during the war. These seem to be by manufacturer rather than intended patterns. The early war wings are as follows: a wing on a half circle of serge with two layers of feathers on each wing. The shrouds of the parachute are "lined and couched" which is to say the piece that is the shroud line is lain on the serge and then secured by a series of stitches. The second early pattern is that which is embroidered on a think serge rectangle. Then there is the more common pattern which is cropped. A special variant was a wing with a black serge backing that was worn by members of the 156th Battalion who were trained in India. Major Powell, CO of B Coy (Commanding Officer of B Company), wore one of these wings into captivity when he was taken prisoner at Arnhem. Officers did have bullion private purchase wings made for their Battle and Service Dress uniforms. More so for the Service Dress.

Jump qualification badge

Introduced on June 17, 1942 for jump qualified troops in a non-airborne role. Meaning they were not in a parachute unit. This badge was initially worn on the upper right sleeve but was moved to the lower right sleeve 6" from the cuff of the battledress blouse. These badges were not worn on the Denison. Members of the Air Landing Brigade who were jump qualified wore both this badge and or the jump wings. Corps and administrative troops who arrived by glider could also wear this badge if so qualified. The requirements for the jump qualification badge were four jumps. Col. Sweeney of the Oxf & Bucks did mention that he had completed the parachute course and wore this badge on his blouse: "I had to make 2 jumps from a balloon and 5 more from an aircraft."

Glider qualification badge

This badge was worn by members of the Air Landing Brigade as well as by the various Divisional troops. It was authorized for wear by those who had completed their glider training and had made a non specific number of flights in a glider. It was worn 6" from the cuff of the right sleeve and was a worsted badge on serge. There is a brass variant that is thought to be of post-war manufacture.

Glider Pilots Brevets

There were two brevets (wings). One for the first pilot and a smaller one for the second pilot (Co pilot). In April of 1942 the first pilots brevet was introduced. Both the 1st pilot and 2nd pilot wore this brevet. Officers were allowed to private purchase bullion brevets for their Battle and Service Dress. All brevets were worn just above the left breast of the smock and battledress blouse pocket.

The official 2nd pilot's brevet was introduced on August 19, 1944. There were hand made 2nd Glider Pilots wings which were worn prior to the official brevet becoming available. These makeshift brevets were 1st pilots wings with a brass ring in the middle and are very rare.

Shown here is the uncut and cropped Jump Qualification badge known to some as the "Light Bulb." It is sewn on heavy serge wool.

The glider qualification badge, worn on the right sleeve, 6" from the cuff. One had to complete a glider training course and fly in a glider to qualify for this badge. This flight was not confined to combat, but rather was issued after one training flight.

The 1st Glider Pilot's brevet. Above is the variant found on both the BD and Denison, and below is the Officer's Service Dress variant made of bullion. (As a reference, the 1st Pilot is the main pilot, and the 2nd Pilot is the co-pilot.)

Brass glider badge. The author has never seen one on war time BD. This may be a post-war variant. The embroidered glider badge is the only variant the author has seen worn on BD, in both wartime photographs and on physical examples.

Chapter Seven: Airborne Insignia

Clockwise are: the printed shoulder title (the final pattern to be manufactured, made in both printed and embroidered), Officers Airborne piped crown, plastic Army Air Corps badge, 2nd Glider Pilot's brevet, and the 1st pattern printed shoulder title. Note this title was also worn by the Paratroops in the very beginnings. These printed Airborne titles are uncut and as issued. The soldier would cut them out along the slotted lines and then sew on to his BD. Also of interest is that the Parachute Regiment title was spelt out and the Glider Pilot abbreviated. Note that the 2nd Pilots brevet was not issued until 1944 – until then both 1st and 2nd Pilots wore the 1st Pilot's brevet. The distinction between the 1st and 2nd Pilots was in name and for their position in the glider, and for the duties they performed.

Right: Harry prior to his promotion to Staff Sergeant. Note the lack of brass crowns on his blouse – thus he held the rank of Sergeant at the time this photo was taken.

Flashes and Strips

Pegasus Flash

This flash was introduced in May of 1942. Designed by a Major E. Seago and based on the Greek heroic saga of the winged horse Pegasus and his rider Bellerophon. There were two variations, one embroidered and one printed (wartime economy). Those printed were less costly to make, as every little effort helped the war effort. Printed flashes were screen printed onto cotton twill, using a reverse negative. Meaning the Pegasus would be sky blue, which was the color of the material. Maroon ink was screened onto the blue, forming the outline of the Pegasus.

A soldier could be issued with either embroidered or printed, though for the war effort printed was more likely to be found on the parade (drill) and or for combat blouse. Embroidered were reserved more so for the best dress blouse. Printed flashes were issued in one long rectangular piece, being that the right and left were on one piece of material – as they were printed. The soldier would then cut the rectangle in half-splitting the left and right Pegasus, 1/16" to 1/4" was folded underneath to prevent fraying and it was then sewn to the blouse. Both Officers and other ranks wore printed and worsted. Silk embroidered flashes were made available to officers for their service dress.

Right above: The front and back of the printed Pegasus. Note the dotted lines which are used as a guide for folding as printed flashes must be folded over and then sewn to the BD, otherwise the flash will fray with wear. Some examples in the author's collection were cut on the dotted lines and then folded and sewn making a smaller square flash. This shows that there are exceptions to all regulations.

Right: Grouping of variants. Note that the cotton twill is blue and the maroon ink is then spread over a reverse negative screen.

107

For King and Country: British Airborne Uniforms, Insignia & Equipment in World War II

The printed and embroidered shown side by side. Note that the flashes are cut differently – the worsted rectangle shown here was more commonly cut into squares.

Note the variations in detail and manufacture of these embroidered Pegasus flashes.

Reverse of each embroidered Pegasus flash.

Clockwise: printed flash folded on the dotted line, smaller printed flash folded on dotted line, uncut as issued embroidered pair of flashes, printed flash with khaki backing, and two variants of embroidered flashes. Note the two blue starting points on the uncut pair. This is to aid in cutting them in the exact middle.

The reverse of the previous grouping. Note how the top and bottom of the printed flashes are folded first, then the sides are put under an iron and are ready to be sewn on the BD. Also of interest is the printed flash backed with a piece of khaki twill with brass hooks added, allowing it to be removed for cleaning – this is most unusual.

108

Chapter Seven: Airborne Insignia

AIRBORNE Strip

This strip was to replace the AIRBORNE title in 1942 for wear by non-Parachute troops. Yet it was worn by British parachute troops until December 31, 1944 when it was sternly discontinued. Both printed and embroidered were manufactured. The lettering for the embroidered strip was very different from the type set used on the printed strips. Glider Pilots, Airlanding Brigade and Divisional troops wore the strips throughout the war. Of interest is that the printed Airborne strips were screen printed on cotton twill with a reverse negative. Col. Sweeney of the 2nd Battalion Oxf & Bucks commented that they did away with the AIRBORNE strip in early 1943 (see his foreword, the photograph of the Col. and Field Marshall Montgomery. The Col. is not wearing the AIRBORNE strip. This is simply a recollection of the Colonel's and applies to his unit only.).

War Correspondents flash

War Correspondents went into combat to report on the various operations throughout the war. A special slip on flash was worn by these men over the shoulder straps of their Denisons and Battledress. These flashes were dark green with embroidered gold letters 'BRITISH WAR CORRESPONDENT'. Three such reporters who went to Arnhem were Alan Wood, Jack Smyth, and Stanley Maxted. Non-British correspondents, such as Polish correspondent Marek Swiecicki, wore a similar flash with these gold letters. 'ALLIED WAR CORRESPONDENT'.

Army Film and Photographic Unit flash

Men of the A.F.P.U. went into combat to shoot (photograph and film) the various operations. Those who were jump qualified parachutists wore the jump wings on their Denison and the Parachute Regiment cap badge on their maroon beret. The special embroidered flash was of the red letters A.F.P.U. with a white camera on a black flash and worn on both the smock and BD blouse. The following A.F.P.U. men were at Arnhem: Sergeant M. Lewis, Sergeant Gordon "Jock" Walker, and D. Smith.

Shoulder Titles

There are many variations that were worn by the Airborne. Most were both manufactured in printed and embroidered versions. Embroidered was usually reserved for the 'Best Dress' or 'Walking Out' blouse. Printed is more commonly seen in use on Parade and 'Battle' Blouses. This was not the rule or regulation.

Two printed Airborne strip variants.

Here can be seen how the printed Airborne strip is folded and then ready to be sewn on the BD.

The 1st pattern printed Airborne title. As mentioned previously, titles were cut along the slotted lines and then sewn onto the BD. Printed titles are different than printed flashes as they cannot be folded over the edges of an arched title. Therefore an extra layer of twill was glued to the title to reinforce it and then it was sewn on.

A 'Badged up' blouse, meaning that the badges are good and have been added to the blouse in hopes of passing as an original. Besides the shoddy sewing job the thread used to sew the badges can be a dead giveaway such as the use of nylon thread which was not in use at the time.

For King and Country: British Airborne Uniforms, Insignia & Equipment in World War II

Note that will I use the British description of for example 'white on scarlet' to describe the colors used on a shoulder title or flash. This means that this particular title features white lettering on a scarlet background.

The Parachute Regiment

In general, worn in this order: AIRBORNE was introduced in May of 1942 along with the Pegasus flash and discontinued by late 1942. It was originally intended for non-parachute troops. This was replaced by the 'Parachute' title. Note that General Urquhart wore the AIRBORNE title into 1944.

The PARACHUTE title was intended for the Parachute Battalions and was first introduced in late 1942 and worn until 1943 when it was then fazed out by the Parachute Regiment title. Exceptions being that the 1st, 2nd, 3rd, and 4th Parachute Battalions wore the PARACHUTE title with their battalion numbers (1, 2, 3, and 4) underneath. These are very rare. These were seen worn by the 2nd Battalion as late as Arnhem. The 21st and 22nd IPC wore the PARACHUTE title with their Roman numerals sewn underneath like the 1st through 4th Battalions. It is known that members of the 21st IPC were ordered for security reasons to remove the number tabs prior to leaving for Arnhem. This would leave simply the PARACHUTE title on their blouses. This title was replaced by the title ARMY AIR CORPS in mid 1943.

As both the Paratroops and Glider pilots were part of the Army Air Corps they wore this title ARMY AIR CORPS until the end of 1943 and was replaced by the title PARACHUTE REGIMENT. This was when the Parachute Regiment and Glider Pilot Regiment became their own entity and were proclaimed official regiments.

PARACHUTE REGIMENT was introduced in 1943 and worn for the remainder of the war. Note that the lettering was dark blue on light blue and not the blue on maroon which is post-war.

The Airlanding Brigade

The Air Landing Brigade wore a wide array of official and unofficial titles on their BD. The British Army High Command had hoped that the Air Landing troops would conform by wearing the same titles as their Line Infantry Parent Regiments which were white on scarlet, or black on rifle green for Rifles for the likes of the Royal Ulster Rifles. Some troops did conform yet many did not and continued to wear their 'special' unofficial titles almost defiantly as the mark of the old salts. Here you will see the unofficial titles that were commonly worn by these Battalions.

The 7th Kings Own Scottish Borderers did not wear shoulder titles. Instead they wore their clan tartan, Leslie Tartan. This is a square flash that was positioned at the top of the shoulder. Beneath this was the Pegasus and then the Airborne strip.

The 2nd South Staffords wore three variants. A white on scarlet SOUTH STAFFORD, a white on scarlet S. STAFFORD, and a yellow on dark cherry S. STAFFORDS in serifs. This yellow on dark cherry being the unofficial title.

The 1st Border preferred their original regimental colors of yellow on green edged in purple to the white on scarlet title.

The 2nd Oxf & Bucks preferred their unofficial FIFTY SECOND title to the white on scarlet. When questioned what title did you wear, the first and most prompt reply from 2nd Battalion Oxf and Bucks Tish Rayner was "FIFTY SECOND!" Did you ever wear anything else besides that? "Nope, FIFTY SECOND!" I've found that most veterans do not remember such details about uniforms or insignia as they had more important things on their mind like staying alive. Though it is hard to dispute such a strong reply as Mr. Rayners, and mind you this was without any sort of prompting.

The printed PARACHUTE REGIMENT above, and the embroidered below. Note how wear has stretched the embroidered title out of shape. The title PARACHUTE REGIMENT was issued on August 1, 1942, yet not officially authorized until August 31, 1942.

Copies of the 21st independent Parachute Company. Note that this is a two piece badge. Prior to Arnhem the troopers removed the XXI for security reasons. Also note that this is the same PARACHUTE title that replaced the AIRBORNE title. Both the PARACHUTE and AIRBORNE titles were issued in printed and embroidered.

Right: The 22nd Independent Parachute Companies title. The author is aware that only the 21st and 22nd were issued embroidered and not printed. However, only the PARACHUTE was printed, and the XXI and XXII were not.

110

Chapter Seven: Airborne Insignia

Col. Sweeney replied to this with the following: "I wore the FIFTY SECOND titles as all others did until late 1942, when we were told we had to wear the new white on scarlet title as our mail was being sent to the FIFTY SECOND Anti-Aircraft Battery who were also part of Southern Command. Though I do think it was rather a matter of simplicity that we changed our titles to the white on scarlet OXF. & BUCKS as it was cheaper to reproduce and every little bit counted." In a photo with the foreword of this book, the Colonel is being decorated by Montgomery – note that he is wearing the white on scarlet title. What is even more interesting is that Tish Rayner of the same Battalion claims he only wore the FIFTY SECOND title. Both men can be correct, as some members of the Battalion did not exchange their old titles for the newer variants. This goes to reinforce my opinion that exceptions are indeed possible. As the 12th Devons wore the white on scarlet it is very possible that they did wear both printed and worsted titles.

The 1st Royal Ulster Rifles wore their unofficial blue on black title.

The Air Landing Brigade did not as a rule wear printed titles but rather worsted. These were worn in conjunction with the Pegasus and Airborne tabs which were both printed and worsted. Examples will be shown with their corresponding cap badges.

The Glider Pilot Regiment

Worn in this order: AIRBORNE (the curved title) from February 1942 to the Fall of 1942) until they adopted the Army Air Corps title.

ARMY AIR CORPS was worn by the early Glider Pilots, until early 1943 when they adopted the new GLIDER PILOT REGIMENT title when they were made their own regiment. The GLIDER PILOT REGIMENT title was worn from this point and on through the war.

Corps and administrative departments

Corps and administrative departments wore their individual titles in conjunction either with or without the jump wings, Pegasus and airborne strip as seen here. Cap badges shown are plastic economy, bimetal and brass. Of note is that even the Corps had unofficial titles as the Airlanding and Paratroops did. Two examples are: the title ROYAL SIGNALS was in fact the official war time title, as the ROYAL CORPS OF SIGNALS was unofficial. The R.A.M.C. was the official, as there was a non-official title of gold serif lettering on maroon ROYAL ARMY MEDICAL CORPS.

Shoulder Titles of the Glider Battalions seem to be predominately embroidered, though further investigation may prove otherwise. Note that this is a copy, as originals of the "Unofficial Titles' are next to unobtainable.

Two of the three title variants worn by the Glider Pilot Regiment. The first variant was the AIRBORNE, then the ARMY AIR CORPS, and lastly the GLIDER PILOT REGIMENT. The GLIDER PILOT REGIMENT title was issued on the 24th of February and officially authorized on August 31, 1942. All three were made in both printed and embroidered versions.

The economy insignia of the Glider Pilot Regiment. Clockwise from the right: the plastic A.A.C. cap badge – notice the brass keeper behind the bird's tail feathers. These keepers were stuck through small slits made in the beret, and once the keepers were passed through they were folded over to secure the badge. Below the badge is the printed shoulder title – note how "regiment" was not spelled out but rather abbreviated. Lastly, the printed airborne titles are an uncut, as issued, pair. The soldier would cut these on his own and then sewn them onto his blouse.

The Royal Army Chaplains Department embroidered title and grouping. Of interest is that the soldier's bible belonged to the author's great uncle who served in World War I and who carried this with him for comfort. There was an unofficial title for the chaplains – this was spelt out ROYAL ARMY CHAPLAINS. A printed variant of R.A.Ch.D. was also worn.

G.Q. Keep sake badge

Gregory and Quilter manufactured X type parachutes for the Armed forces. As a special "keepsake" they made a badge to be given to each soldier who completed his jump training course. Upon the wings parade, where one is presented with his wings, the GQ badge would also be given to the individual. This was only done in the early years of the war up to 1943. As the war drew on, this keepsake badge was discontinued. Irvin Parachute Company also issued a similar keepsake badge. Across the front would have been the letters IPC.

Caterpillar Badge

This was awarded to those who made an emergency jump from an aircraft. CSM (Color Sergeant Major) Gatlant of the 11th Battalion wore one such badge. This is truly a rare badge.

Badges of Rank

Officers

Officers wore a series of rank badges and they consisted of pips, crowns, and the sword and baton for generals. Pips were made in brass, embroidered and plastic and were worn 1/2" above the seam that joins the shoulder strap to the sleeve. Each additional pip was worn above the next, side by side positioned like diamonds. When worn on the smock they could be either sewn to the smock's shoulder straps or what is known as a "slip on" would be used. This is an extra piece of material, usually khaki or olive, where the worsted pips would be sewn to this "slip on" and then slide through and over the smocks shoulder strap. The same applies for crowns which also were made of brass, plastic, and embroidered.

Officers wore the following for these ranks:

One pips for Subaltern (2nd Lt.), two pips for Lieutenant, and three pips for a Captain.
Crowns were worn for the rank of Major. One crown on each shoulder strap.
Lieutenant Colonel wore one pip with crown.
Colonel wore two pips and with crown.
Brigadier wore three pips and with crown.
Major General wore a crossed sword and baton with one pip.
Lieutenant General wore a crossed sword and baton with one crown.
General wore a crossed sword and baton with one pip and one crown.

Parachute Regiment and Air Landing Brigade embroidered pips and crowns were white on sky blue, an exception being rifle regiments who wore black on rifle green pips and crowns (RUR for instance). Divisional troops and Administrative Department's pips were backed in color of their branch. Dark blue for Signals and Engineers, yellow for Armoured Corps, purple for chaplains, and maroon for R.A.M.C. for example.

Brass pips and crowns were authorized for wear on Service Dress and embroidered were authorized for wear to be sewn on to BD and Denison. Of note, selected officers of the R.A.M.C. were known to have worn brass pips upon their Denisons at Arnhem, though this is very unusual.

A copy of a very rare title, the Royal Artillery Anti Tank. (I would prefer not to have any copies in the book, yet finding these titles can take a lifetime as they are ultra rare. I thought it important to show you at least an example of what they looked like – Author).

Left: Here is an interesting piece. This is the little keepsake badge presented to newly qualified parachutists by the parachute manufacturer Gregory and Quilter. It was Raymond Quilter who invented the British made quick release box. Right: Stamped into the reverse are the letters G.Q. and the word PARACHUTIST. The author has another example that is unstamped and plain on the reverse.

Two variant pair of Officer's Airborne worsted pips. Each pip is raised, meaning that it is filled with cotton to make it 'bubble'. Pips are worn as seen here across and on the tip like a diamond. The differences in color are simply a variation in manufacture.

Chapter Seven: Airborne Insignia

Here a patient Corporal waits for his shot. Note the worsted rank chevron worn on the sleeve of his smock (worsted were common issue for the European Theater of Operations). Note there were printed chevrons though these were not, as far as the author has seen, worn by the Airborne or E.T.O. based line infantry for that matter. As a note, the term worsted means simply that that particular badge was sewn to/on thick serge wool. It also applies to badges of rank, such as pips and chevrons.

Other Ranks

Rank Chevrons (stripes) were worn predominately on the right sleeve of the Denison for Paratroops, and on both sleeves of the Battledress. The Air Landing Brigade seems to have worn them on both sleeves of both the smock and BD. Glider Pilots predominately wore their chevrons or chevrons and crowns for a Staff Sgt. on both sleeves of their smocks. Evidence supports that some did only wear their rank on just the right sleeve, though this is rare.

Rank Chevrons were worn as follows: Lance Corporal 9" from lower point of chevron to the top of the sleeve. Corporal 9 1/2" from lower point of chevron to the top of the sleeve. Sergeant 10 1/2" from lower point of chevron to the top of the sleeve.

Warrant Officers wore their badges of rank 6" from the cuff of their Battledress blouse. I have not seen a warrant officer's badge on a Denison, though I am sure it is very possible. There was a crown within a wreath for WOII (Warrant Officer 2nd class), and the Royal coat of arms for a WOI (Warrant Officer 1st class).

Chapter VIII

Pattern 37 Webbing

Called Pattern 37 as this is the year it 'Came to be'. Each piece is stamped with the manufacturer and year of manufacture, and a war department stamp usually in the form of a WD and 'broad arrow'. Brass buckles and press snaps are found on wartime webbing. British webbing is of a khaki/tan shade, with Canadian more of a golden color. British and Canadian webbing were the best made in durability and quality. Indian, South African, and Australian were not as well made and would not be used by the Airborne. Both the British and Canadians made their own variants of each piece of webbing.

Web Waist-Belt
Waist belts came in four sizes: 'EL' extra large, 'L' for large, 'N' for normal, and 'S' for small.

Web Braces
Braces came in three sizes: SMALL, NORMAL, and LONG. Sometimes referred to as cross straps, braces come in pairs, left and right. One brace has a loop for other brace to slip through. This keeps the braces better situated on the body. From the front the braces fit underneath the belt, up through the slot in the basic pouch through the brass keeper and then over the shoulder where they crossed at the back and lastly attached to the back of the waist belt. Underneath is the piece description, being either 'SMALL', 'NORMAL', or 'LONG'. Longs were issued to the taller soldier as he could not assemble his kit properly with shorter braces.

Basic Pouch MK II
Used throughout the war, it was shorter than the MK III and would not hold Sten magazines. This led to the manufacture of the MK III.

Basic Pouch MK III
One inch taller than the MK II. It was made specially to fit and carry Sten magazines. Both the MK II and the MK III had a brass buckle sewn to the top back of each pouch. This was for the attachment of the web braces and also for the brass hook of the L-strap which is attached to the haversack. Sten magazines would rattle about inside which led to the manufacture of the Sten pouch, which had proper slotted magazine dividers inside. The MK III could carry four to five (if you were extremely fervent) magazines. For the basic rifleman, this pouch was designed to carry: two 2" mortar bombs, four .36 mills bombs or 4 Bren magazines (two magazines per pouch). Rifle ammunition was not carried in this pouch, rather in a canvas bandoleer which was worn around the neck and shoulder.

The 'Skeleton Order' of P-37 webbing – called skeleton as it was as light as the soldier was wanted to be in the field, this being without haversack and large pack. Clockwise are: one pair or braces known as shoulder straps, basic pouch MK III, water bottle carrier, entrenching tool cover, bayonet frog, basic pouch MK III and the waist belt. Note the .303 bandoleer, as a soldier's ammunition was carried in a bandoleer and not in the basic pouches.

The LONG brace which was made with the tall soldier in mind. In the middle is the waist belt and below that the E Tool cover. Note the stamping on the brace: manufacturer, date, Broad Arrow and size. On the waist belt is the stamping: S for small, manufacturer, and date. The E tool cover was stamped on the inside.

Chapter Eight: Pattern 37 Webbing

Shown here is a blancoed waist belt. The trouser braces are of civilian manufacture.

The water bottle carrier and basic pouch MK III. Note the string that secures the cork to the water bottle, and the brass snaps found on the webbing. These are on nearly 99% of all pouches and pieces. Note that the basic pouches dated 1944 with the toggle and loop closures were not worn in the E.T.O., or more so, by the Airborne in the E.T.O.

The LONG braces. Note how they go under the shoulder straps and fit into the brass buckles of the waist belt.

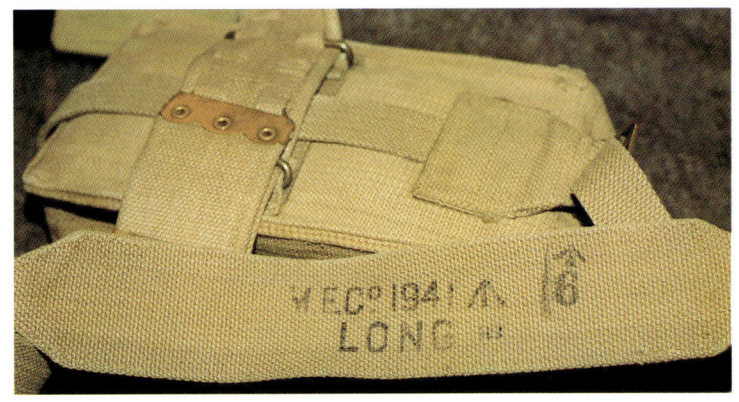

A LONG brace which is fit into the basic pouch and then underneath the waist belt. Also note the brass hooks that fit into and secure the pouch to the waist belt. M.E.C. was a common manufacturer of webbing during World War II.

Note how the brass hooks of the Basic Pouch fit into the waist belt, and how the shoulder brace comes down under the waistbelt and then over to hold up the water bottle carrier. The other side of the water bottle carrier is attached to the other brace which is then fed through the E Tool carrier and waist belt. The MK III Light Weight respirator is attached to the waist belt with the same brass hooks as found on the basic pouches.

115

Anklets

Here you see two variants, above has the web straps and a canvas inner guard. This is to better preserve the piece against wear from friction from the boot. Below has the leather straps and leather inner guards. What is interesting is that the leather is that of the war effort grade often used to make Home Guard equipment. Each would be sized by number. Sizes ranged from 1 to 4 – size 3 being normal and 4 being an extra large.

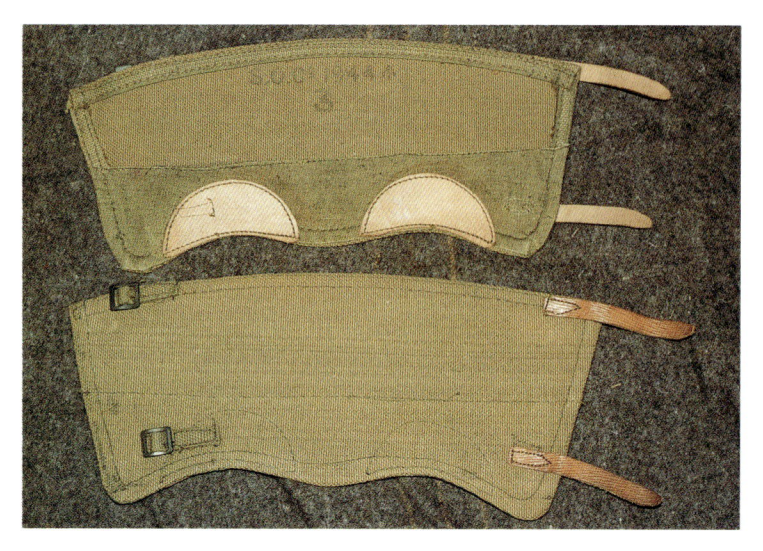

Variants of web anklets which have leather straps – most common are those with web straps. There are 4 sizes. Note the stamping: manufacturer, date, broad arrow, and size.

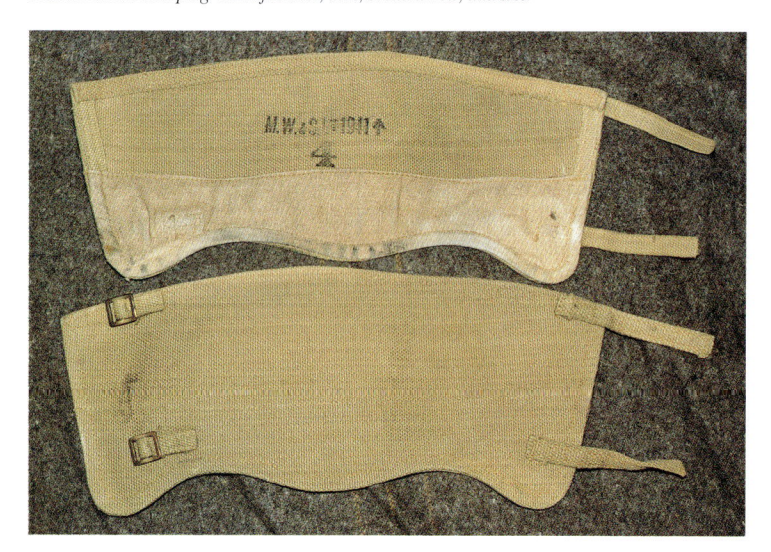

The most common variant of anklet. Note the stamping: manufacture, date, broad arrow, and size. Size 4 was the largest.

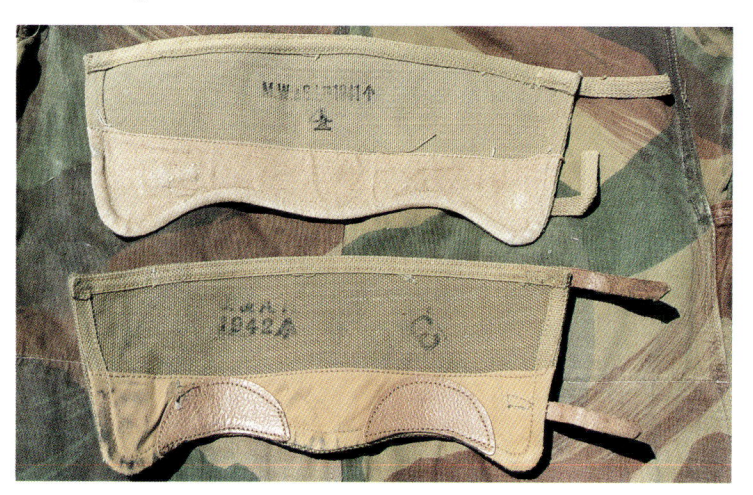

The MK III Basic pouch with brass closure snap. Above the pouch is the hook of the L-Strap.

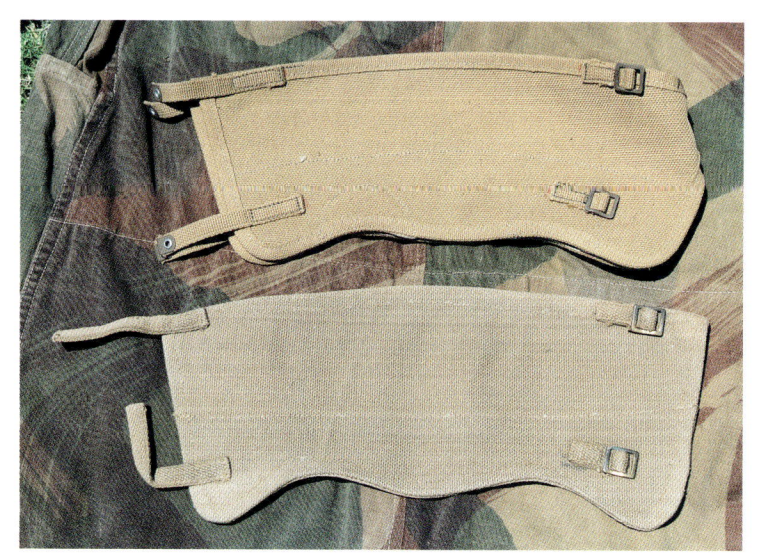

Above is the Canadian anklet. Note the color difference between the Canadian and British made shown below. Canadian webbing is all of this golden color. Most of their buckles and hardware are painted with a khaki color, as the British are bare brass.

Left: For comparison, the two variants.

Chapter Eight: Pattern 37 Webbing

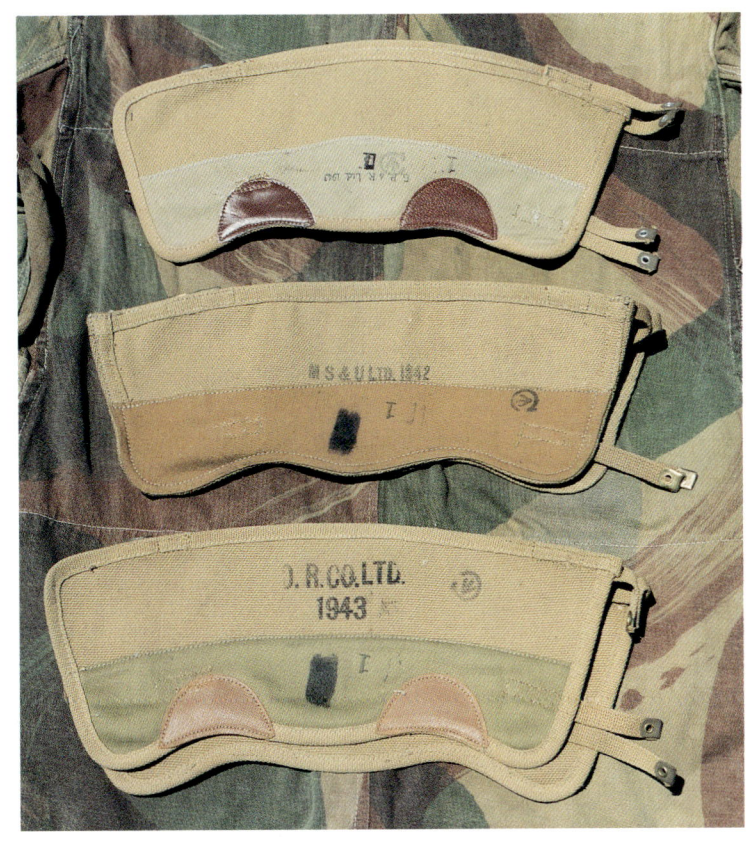

Three Canadian variants. Note that all Canadian webbing is easily identified not only by the golden color, but also the stamping. Each piece of Canadian webbing has a 'C' broad arrow – note the 'C' around each broad arrow.

Large Pack

Used by Signals, RAMC and the Parachute Squadron Engineers, which were marked "E" for Engineers. These were used to carry special equipment and or supplies, and not personal items as the Line Infantry would carry in theirs. There is a photo of Glider CRA, where Thompson CRA is carrying a large pack. When asked why does an officer carry his own pack, he replied, "my bottles of whiskey were in it." As a rule, Large Packs were not used by the Airborne infantry.

Water bottle web carrier

There are two variants, one, the most common and preferred, was the skeleton carrier, and secondly the sleeve carrier. The latter was not popular as it was difficult to get the water bottle out of and then back into the carrier. Either style was worn to the side on either hip.

Above: Shown on the PIAT Gunner's hip is the skeleton water bottle carrier. Note that this soldier's webbing is unblancoed.

Left: On the rifleman's right hip is the Sleeve water bottle carrier which was not very popular as it was difficult to get the bottle out and then back in, not to mention under fire!

Entrenching Tool

This was the soldiers personal entrenching tool. Sections did carry full sized picks and shovels, usually one per section (10 man squad). The Entrenching Tool (E tool) is broken down into three pieces. One the tool/blade, which worked both as a pick to break ground and a shovel to scoop out the broken earth. A wooden helve handle which came in two variants. The first variant was a solid piece of wood. The second variant featured a bayonet lug for the spike bayonet. When the bayonet was attached to the helve handle it was used for mine probing and even as a weapon in desperate situations. And third was the web carrier. There were two variants. One a simple tool envelope carrier with web loops to hold the helve handle. There was a drawback to this first pattern carrier, that is the helve handle would fall out, rendering the whole piece useless if it were lost. The second variant had an extra strap added to prevent the handle from sliding out. As far as the author knows this was done no earlier than 1943.

The 'E' Tool was carried on the lower back, though the author has seen it worn on the right hip – this was mostly done by Canadian Infantry (not Airborne).

117

For King and Country: British Airborne Uniforms, Insignia & Equipment in World War II

The inside of the E Tool carrier. Note that it is lined with a heavy twill. The stamping will be towards the inside opening. The tip of the tool inside can just be seen. Note how the brace is threaded to hold both the water bottle and E Tool carriers.

Here can be seen the Entrenching Tool Carrier. The tool is just visible where the carrier is folded over partially – this is the common variant. Note that no water bottle carrier is worn, so this Para must have his bottle in his Haversack. Note that this is where it was originally intended to be carried, as there was to be no webbing below the waist.

On the left is the late war E Tool Handle with spike bayonet attachment. In the middle, and in the right are the early war E Tool handles which do not have bayonet attachments.

Right: Shown here is an unusual E Tool carrier variant. Note the extra piece of web strap on the end securing the E Tool handle. Called by some the Airborne model, it is simply a variant, and an uncommon one at that. (though smart as the handles were known to slide out.) Note that the water bottle carrier strap is left undone as was a common practice in combat. Also note that the chin strap is left loose, another common practice.

118

Chapter Eight: Pattern 37 Webbing

Web Bayonet Frog

There are three variants. One for the sword bayonet which accompanied the No.1 Mk III Rifle. This rifle and bayonet were used for the first two raids and then discontinued. The No.4 MK I was issued to the Airborne just prior to the landings in North Africa. The second variant frog was made to fit either the sword or spike bayonet. The third variant was made only for the spike bayonet. The frogs were worn on the left hip. Each had a loop which slid over the waist belt. Loops and slits in the webbing secure the bayonets scabbard.

Brace Attachments

Designed so that the wearer could wear his webbing without the use of basic pouches. On a normal webbing 'rig' the pouches must be included. The attachments were for Officers and specialists, enabling them to wear their special pieces of kit. Officers, officially were to wear a belt, braces, brace attachments, compass pouch, holster and pistol ammo pouch, map case, and binocular case. As they were Airborne, variations were made according to the personal needs of the soldier.

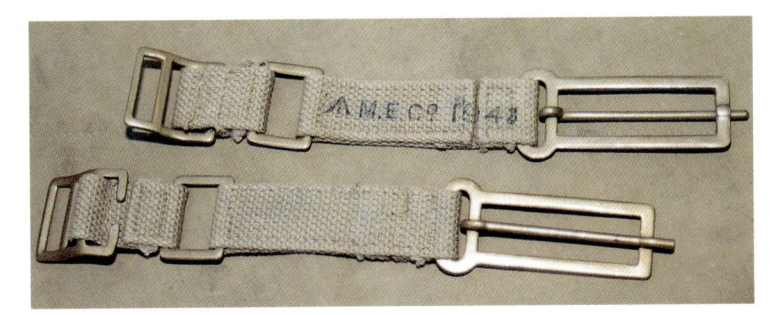

The brace attachments. Stamping denotes: broad arrow, manufacturer, and date. The waist belt was folded over and slipped through the rectangle brass opening with the brass pin fitting behind the belt, thus securing the belt to the brace attachment. The buckles on the left (top when worn vertical on the webbing) were for the web brace to fit through. There is another brace variant issued in 1944 which was a simple semi-open buckle.

One of the three bayonet frog variants. This type fits only the spike bayonet. Note the slit for the scabbard nub to fit through thus making it secure. Also note the Austerity Pattern trouser map pocket.

Compass pouch

Carried by officers, NCOs and specialists. Padded with felt to protect the compass.

Left: The compass pouch. Note the stamping: manufacturer, broad arrow, and year. The compass is a Prismatic and very collectible. Just visible is the top of the padded lining. Right: The markings on the Prismatic compass: manufacturer, broad arrow, year and type.

For King and Country: British Airborne Uniforms, Insignia & Equipment in World War II

Above: The reverse of the MK II compass. Note the markings: type, broad arrow, and manufacturer.

Right: Reverse of the compass pouch. Note the brass hooks which fit into the waist belt to secure the pouch. The web loop on the bottom of the pouch is for the binoculars case. The web brace fits through the top loop and then through the brace extension which is attached to the waist belt. The binocular's case had a pair of brass hooks that fit into the lower loop of the compass pouch. Note the Mk II compass on the left of the Prismatic.

Holster, Webley

There are three sizes, one for the standard revolver which held the .32 Colt Automatic and the No.2 revolver. Another variant was 1" longer and designed to accommodate either the Engils Hipower or the U.S. made .45 Automatic as well as the No.2 revolver. A third holster being for the larger .455 Webley revolver, though the .455 was not issued to the Airborne.

Pistol Ammo pouch

Carried by officers, NCOs, Bren gunners and specialists. Same as the compass pouch, though void of the felt padding.

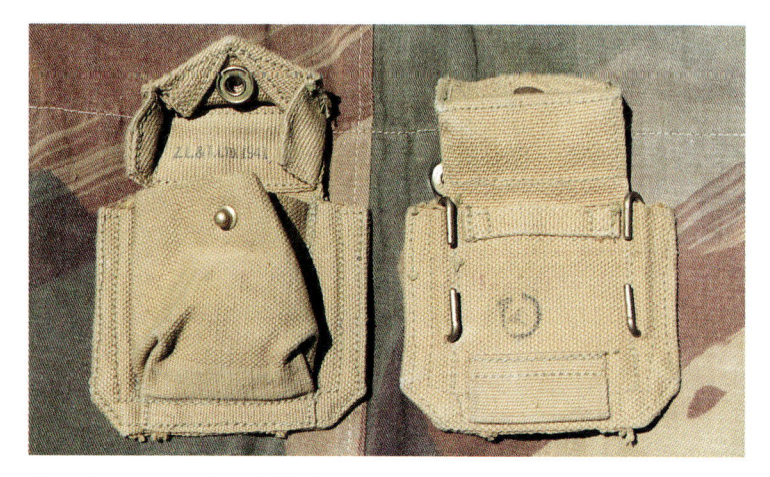

The revolver ammo pouch. On the left is the British made, and on the right the Canadian. Note the loop on the bottom which was for the attachment of the web holster.

Wire cutters pouch

Made for both the large and smaller wire cutters. To my knowledge there are no Canadian variants of this piece.

Binocular case

Padded on the bottom for added protection. Carried by officers and specialists.

The binoculars canvas case. The case is reinforced on the sides and thickly padded on the bottom for the protection of the binoculars. Note the web strap, as leather would rot in the field.

120

Chapter Eight: Pattern 37 Webbing

Haversack (Small Pack)

Issued to and carried by both the Para and Glidertroops. As these soldiers were designed to be 'shock' troops, the contents of the haversack was limited and condensed for mobility's sake. They were designed to be worn and carried on the back, though members of the Airlanding Brigade wore them to their sides attached to the web braces. I can only guess that this is because they had to sit with their backs to the inside of the glider and this could not be done if they were wearing haversacks. When asked if they wore their haversacks upon their backs or to their sides without L-straps Col. Sweeney of the 2nd Battalion Oxf and Bucks replied, "I certainly can remember wearing it both ways but I'm not certain if we had a special drill for later attaching the L-straps or not."

When worn on the back a pair of L-straps were attached to the haversack. These were the shoulder straps. Attached by a set of brass buckles. These straps could be easily removed if desired. The Haversack had three compartments, one for the Mess tin with rations inside. A second pocket was initially designed to carry the water bottle as no equipment was to be carried below the man's waist, thus the water bottle carrier was designed later leaving this extra compartment in the pack. Some troops did carry an extra bottle in this pocket. The third and larger of the compartments was for the ground sheet and other personal kit. See the personal items section for a complete list

Left: Shown here is the 'packed' haversack for a typical operation. One L-Strap per side with its brass hook to fit into the top loop of the basic pouch. Of interest is how the tea mug is carried – this was very common yet in combat it was sometimes placed inside of the haversack. Also note the ground sheet as carried by regulation. Right: The inside flap of the haversack. Note the stamping: manufacturer, date, and broad arrow. The ground sheet is folded over the front.

Here the Haversack worn as per regulation and filled with the soldier's personal items.

For King and Country: British Airborne Uniforms, Insignia & Equipment in World War II

Above: Canadian Haversack with matching L-Straps.

Above right: The reverse of the Canadian haversack and L-Straps. Note that the British version is identical in construction yet not in color. The twin hooks that fit through the loops found on each basic pouch keep the haversack in place.

Right: The brass L-Strap hook which was fitted into the basic pouch upper loop. Of interest is the loose fitting sleeve and wrist tab button of the 2nd pattern smock, and just visible the ground sheet hanging from behind the haversack.

Chapter Eight: Pattern 37 Webbing

Satchel, Officers

The officer's satchel featured two large and two smaller internal compartments. A shoulder strap was attached to each side by a brass buckle.

Above: The Officer's Satchel. Note the brace attachment's brass buckle that is fit over the waist belt.

Above right: A Canadian Officer's Satchel. Note the special shoulder strap which looks like a common shoulder brace but is padded.

Right: The inside flap of the satchel. Note the C broad arrow. Inside are three pockets.

Bren Auxiliary pouches and yoke

The Aux. pouches were deeper and larger than the MK III pouches and were able to carry three Bren magazines in each. The pouches were joined by a web yoke which was worn around the shoulders and neck. The yoke could be detached and the pouches then attached to the Bergen Rucksack providing it had the special web straps added to its leather shoulder straps.

Left: The Bren auxiliary pouches and shoulder yoke. Note the waist strap which secures the pouches firmly to the body. Three Bren magazines are carried in each pouch as opposed to the MK II or MK III basic pouches that carry two magazines. A Bren Gunner and Assistant Gunner would wear one of these with his basic webbing. As the Bren was the section's support weapon much ammunition was needed to keep up the fire.

For King and Country: British Airborne Uniforms, Insignia & Equipment in World War II

Above: The Bren Aux. pouches with their full load of six magazines.

Right: Pictured here is Section Leader P. McTavish of No.6 Commando loading 2" bombs into his auxiliary pouches. Note that they had many uses and could carry mills bombs, mortar bombs (as shown) and Bren magazines. Note how the holster is attached to the .38 ammo pouch by a pair of brass hooks and web loop. Also of interest is the P-40's buttons which are the earlier variant. This blouse is dated either c. 1941-1942.

Tanker Holster

This was a favorite among the Recce and 'Old Salt's, and was issued in the early years of the war, and later cut and modified, as the long 'gun fighter' leg strap became obsolete. It was later issued in a simplified version, without the long leg strap. This piece was common among the 1st Division especially.

Left: The Tanker's Holster. Designed with the cavalryman in mind these holsters very popular among the Airborne. These holsters were shortened later in the war. Of interest, note how the ammo pouch is attached to the belt and holster, and how the smock's tail is secured to the rear. The butt of the MK II 'S' Sten is also visible. The revolver's cleaning rod is shown carried as part of the holster.

Above: Note how the end metal tip of the holster strap is painted – this is common on much of the Canadian webbing. Also note the C broad arrow on the leg strap. One last detail is the lop on the pistol rod which is for a cleaning patch. The rod is then run through the barrel with the patch absorbing the dirt and cleaning fluids.

124

Chapter Eight: Pattern 37 Webbing

Hipower Holster

Featuring a fold over flap, with a internal magazine pouch, able to hold one 12 round magazine.

Hipower Magazine pouch

One of the most sought after pieces of webbing, and quite unobtainable. This featured twin magazine pouches with a brass snap flap.

Shell Dressing bag

Carried by stretcher bearers, aidmen, and orderlies. This was a simple bag with no internal pockets and a built-in adjustable shoulder strap.

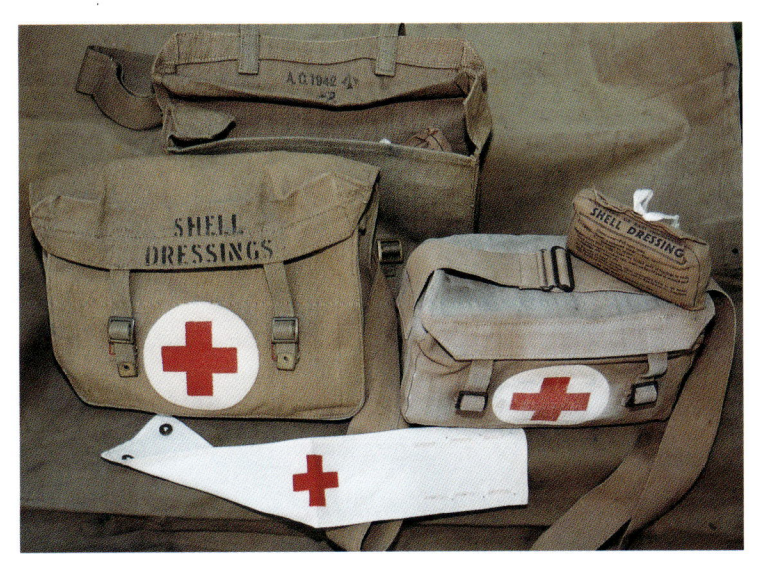

Here can be seen two shell dressing bags and an unknown medical carry bag. On the left note that the contents were stenciled onto the cover, and behind this is an open shell dressing bag of which can be seen the stamping: manufacturer, date and broad arrow. Below is a variant of the aidman's brassard (arm band). Note the three consecutive rows of button holes for the adjustment of the brassard.

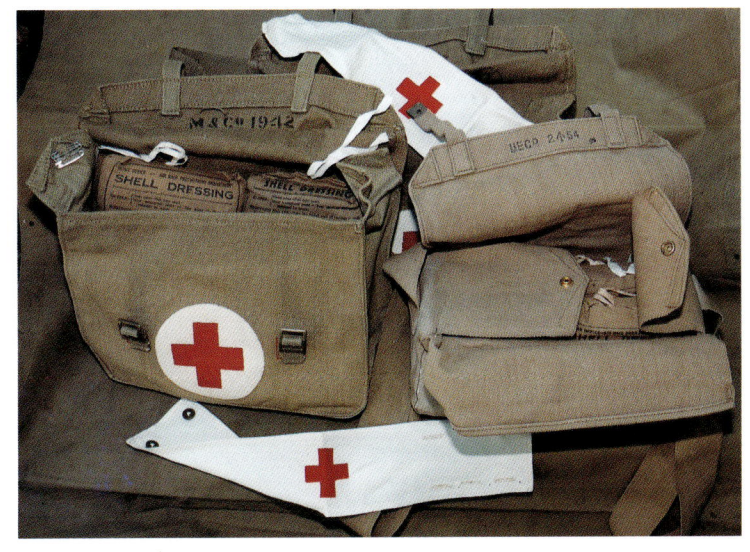

Of interest is the inner flap of the medical bag with securing snap to keep its contents from spilling out. A note on the stamping, MEC is the manufacturer, 2 broad arrow 54 is a manufacturer's code (and not the year of manufacture). Some pieces of webbing found today are not dated yet are stamped very much like what is seen here. The white ties on each dressing are to attach to the shoulder strap to the Denison.

Two Dutch Commandos shown either just prior to, or during, Operation Market Garden. On the left an American jump helmet with web chin strap which was thought not to have been used until the Rhine Crossing – he may be one of the Commandos attached to the American Airborne thus explaining the helmet. The Commando is wearing both the Austerity Pattern blouse and trousers, and note how low he wears his basic pouches. On the right is the hardest piece of P-37 webbing to attain – the Hi Power twin magazine pouch. This Commando wears the Battledress Serge 1940 Pattern blouse and Austerity Pattern trousers with an American trouser belt. Note his Brace Attachment connected to the left shoulder brace, and below to the magazine pouch. Just to the left of the pouch is the Hi Power holster. On his Green Commando Beret he wears the brass Lion of Orange cap badge.

Blanco

Blanco was a powder that was mixed with water and then applied to the webbing to act as a preservant and camouflage coloring. Blanco came in various shades of tan and olive. Olive being the most commonly used in NW Europe.

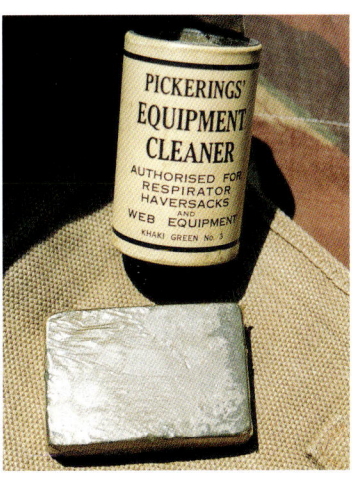

Left: Here is Blanco, one variant, in hockey puck sized tabs individually wrapped in paper. This is the No.3 shade of green. This was mixed with water to a paste-like consistency and applied to the webbing. Right: Here are two other variants of blanco. Note the labeling "Cleaner", blanco was also used to preserve the webbing. The block below is what is known as "D-Day" Blanco which is water proofed rather than water resistant. (Said to be used by assault troops, yet this may simply be another name invented by collectors.)

125

Chapter IX

Personal Items of the Airborne

This chapter will examine what the average Para and glider soldier carried into battle. An Officer's and Glider Pilot's would be similar yet with some additions and or variations. An Officer would have more of the special issue and private purchase items. A Glider Pilot was able to carry more kit in his Bergen rucksack then the issue Haversack. These will be dealt with in a later volume. The items below were also carried in the soldiers pockets as every nook and cranny had to be utilized for troops in the airborne role.

A Paratroop Haversack (is packed for a typical operation, with the following items):

Emergency Ration
The Emergency ration contained a single-block (in 4 chunks) of a high energy glucose chocolate.

24 Hour Ration
The 24 Hour Ration was first issued for the Normandy Invasion. Its contents were held in a waxed cardboard box and lid which fit inside the smaller half of the mess tin. The ration contained:

10 Biscuits.
2 Sweetened Oatmeal Block.
tea/sugar/milk blocks (may be wrapped together).
1 meat block (may be several wrapped together).
2 slabs of raisin chocolate.
1 slab of plain chocolate.
boiled sweets (20).
2 packs of chewing gum (cellophane wrapped solid tabs not strips).
1 packet salt.
6 tablets of meat extract.
4 tablets of sugar.

The suggested menu was broken down like this:

Breakfast:
2 Oatmeal porridge blocks.
2 biscuits.
tea blocks.

What creature comforts the average Airborne soldier might have carried into battle with him. Shown here are the contents of his P-37 Haversack: at the top is his wool pullover, chocolate and boiled sweets tin all of which were loaded with stimulants to keep the trooper awake. Next to the tin is the House Wife sewing kit, Hold All which was the issued toilet kit. jack knife, ground sheet which is an unusual variant as most are khaki in color, mess tin, soap and soap tin, towel, clean pair of socks, foot powder, Bovril which when added to water makes a hot beef broth. There are issued items, and an endless array of private purchase items, and those sent from home.

The Chocolate and Boiled sweets tin. Few tins with insides complete have been seen by collectors.

Chapter Nine: *Personal Items of the Airborne*

During the day:
    2-3 biscuits.
    sweets.
    chewing gum.
    chocolate (to be kept in pockets).

Supper:
    Meat block.
    Biscuits.
    Tea blocks.
    Any sweets, and chocolate left over from dinner.

One knife, one fork, and one spoon

Looking no different than silverware that you might find in grandmothers dining room, these individual pieces were the issue cutlery. Later in the war there was a three-in-one designed after the German variant. The spoon had a slide cover which the knife and fork fit into and clasped together.

One solidified spirit burner.

Known as the "Tommy Cooker." This was the soldier's folding, all in one stove, used to heat his rations and of course tea.

One mess tin

Two deep dished compartments each with a folding handle. One for holding and the brewing of tea, the other for stew and other vittles. The early war model was tin plated. In 1944-45 an all aluminium version was issued. The mess tin was carried in the haversack, in its own pocket compartment.

There is also what is known as a mess tin bag, which in fact is the Unexpired Rations bag. Rations/vittles were carried in this canvas pull string bag, which was then placed inside the mess tin, one lid over the other, thus forming a protective box .

As far as collecting personal items what is seen here are the three hardest pieces to obtain – why? Because the soldiers kept them. There must be countless numbers of the issued cutlery sets in the bottom of granny's cutlery drawer which Granddad brought home from the war.

Tins, Altoids, water purification tablets, which also has a special canvas bag which the water and tablets are mixed in, and cutlery – note the 3 in 1 set copied after the German version. Both the knife and fork are secured by the clamp/hook on the spoon. All pieces of cutlery are stamped with manufacturer, date and broad arrow.

Here is the Mess Tin. A smaller dish fit into the larger dish. Note the folding handles on each piece. These were either cleaned with sand which scoured and scrapped the dish clean or it was left as is until boiling water could be poured onto it.

The stamping on the larger mess tin dish: manufacturer, date and broad arrow. Only the larger of the dishes were stamped.

127

For King and Country: British Airborne Uniforms, Insignia & Equipment in World War II

One towel.
Cleanliness was enforced in the field. These were made of a very thin terry cloth.

One ground sheet
This was a bed n' blanket and rain cover rolled into one. Paras were not issued with rain capes as were infantry regulars, so the ground sheet served as a dual purpose piece. Made of rubberized canvas, it was water resistant. Brass grommets were placed upon the edges so that it could be tied to a tree to make a shelter. Polish troops were issued with the British ground sheet.

One Pullover
The Pullover or Jumper was a 'V' neck sweater that was worn under the BD. No insignia was to be worn on the pullover by other ranks, however there are exceptions of rank chevrons being worn even though it was not regulation. Officers did wear a variant of the sweater with slits in the shoulders so that they could pull their BD shoulder strap through to display their rank. The V neck came in three sizes: sz 1 small, sz 2 medium and sz 3 large. Each had a small cloth size label sewn to the underside of the sweater.

Here is one of several variants of ground sheets. Note the grommets which could be used for a variety of things. Shown here is the heavier, better made sheet as others were thinner and of lower quality. Note that this is how it would be unrolled and then rolled back up and carried in the haversack.

The stamping: manufacturer, date, model (which is unusual to see this), and the broad arrow. The most common stamp is an oval shaped stamp with the usual maker, date and broad arrow. Also of interest is that many do not have grommets, and are simple holes that look as if the grommets were meant to be added yet were not. These are quite common today.

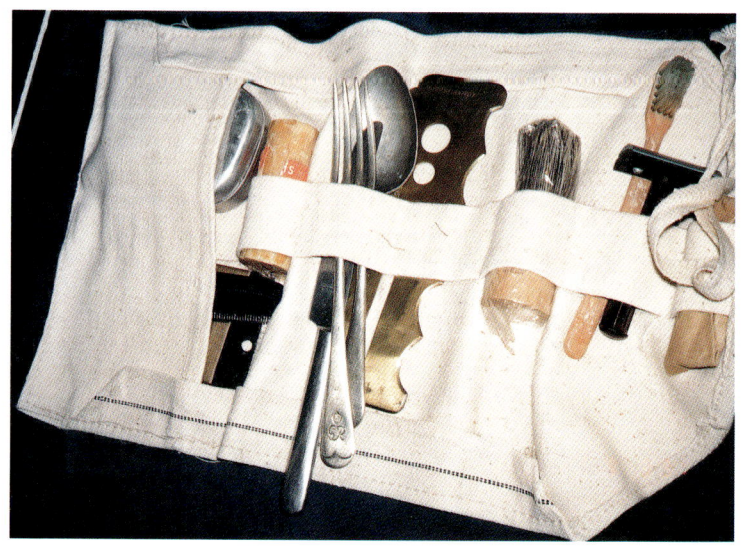

The issue V neck Pullover. Each man carried one in his haversack. It is very common to see them worn in the field, and or by prisoners, for example in Arnhem. These can be difficult to attain, and being wool, are very susceptible to moth damage. Inset: Here is the common size label for the Pullover. The author is aware of three sizes, possibly four. Size 1 is for a man 5' 5", breast 36", waist 28" to 30". These labels are found sewn to the inside of the pullover.

Hold All (Shaving Roll)
Made of heavy linen canvas. This held the following: soap, soap tin, shaving soap and bakelite holder, shaving brush, razor, toothbrush, shoe laces (a spare set), foot powder, and comb. Soldiers were to make a good impression, as they represented the Empire, even while in combat. Therefore cleanliness was mandatory when time and the enemy permitted it. There is recent discussion that the soap tin was not issued until the end of the war as most were dated 1945. Though this is speculation, this does show that new theories continue to emerge.

Tea Mug
Every man had a tea mug! Manufactured in varying colors, the most common were the white with blue rim and the solid chocolate. Carried either in the haversack or hung outside of the haversack. Captain Lorys of the 1st Polish Independent Parachute Brigade said that when they arrived in England in 1940 they were given tea

Hold All containing from left to right: soap tin, comb, mirror, shave stick roll, knife, fork and spoon, brass button polishing stick which was used to keep the brass of the webbing clean and regulation, shaving brush in plastic as would be issued, toothbrush, plastic razor, and razor wrapped in plastic as would be issued. Note the two holes in the button stick. These are for the general list brass button of the service tunic as this stick is of 1917 manufacture, yet was commonly used into the 1950s. Also note that the stamping of the Hold All is usually found either on the underside of the bottom pocket or along the folding sides.

128

Chapter Nine: Personal Items of the Airborne

Above: Issue soap tin and soap which was carried in the Hold All. Most tins were stamped on the outer lid with date and broad arrow.

Right: What a Hold All holds: mirror, toothbrush in plastic as it would be issued, razor wrapped in paper as it would be upon issue, two variants of razor blades, plastic variant razor, shave soap stick which did have a special bakalite carrier, and shave brush. Also in the photograph are blanco and a jack knife with lanyard.

with milk in it: "Tea with milk is for pregnant Mothers and babies." The Poles simply drank it black or with sugar.

House Wife

This was the soldiers personal sewing kit. It was carried in the Hold All. Inside the House Wife were: extra buttons, needles, a thimble, two balls of yarn for darning socks, and an extra piece of Angola wool shirt material which served as a pin cushion.

Jack Knife and Lanyard

Torch

Cigarettes and Cigarettes Tin

Dressings

Gloves

Clothes Brush

Face Veil

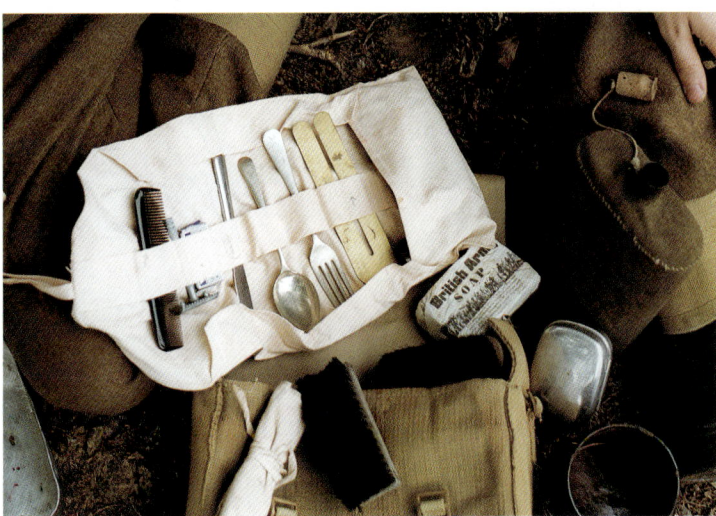

Here a trooper is washing up in the field. Of interest is the brass button stick variant, House Wife sewing kit which is rolled up and held in the bottom pocket of the Hold All, boot brush and clothes brush.

129

For King and Country: British Airborne Uniforms, Insignia & Equipment in World War II

The House Wife sewing kit with angola pin cushion, spool of thread for replacing BD buttons, plastic thimble, and darning wool for the issue socks, of which there were two balls of wool.

The jack knife with marlin spike with lanyard. Most if not all war time knives are stamped with manufacturer, date and broad arrow. The knife features a general blade, can opener, and spike for the untying of knots. This was the most common issue knife.

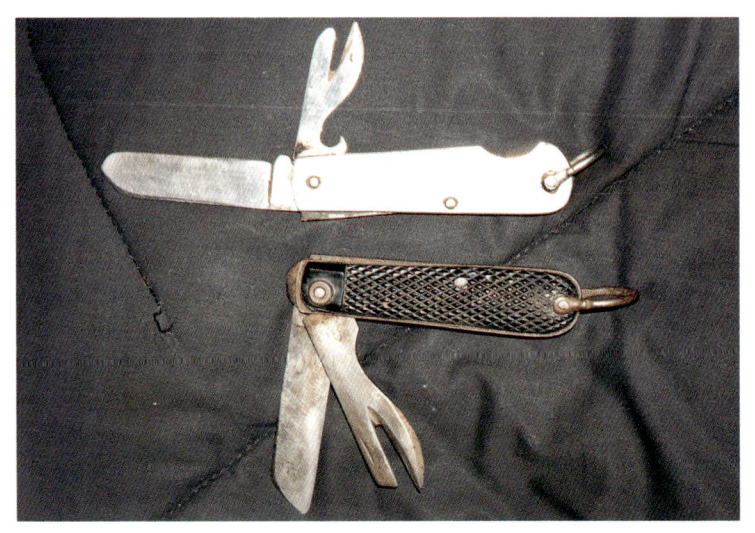

Here are lesser known variants: the all steel was issued in 1945 and did not fare well as they rusted badly. Below this is the jack knife with out marlin spike which features the general blade and can opener. Each have loops for the attachment of a lanyard.

Variations of torches (flash lights) from left to right: copied after the German issue lamp this light can has multi-colored lenses, green and red. Button holes are found on each leather tab allowing them to be worn on the BD or smock. Secondly is the U.S. manufactured TL-122A an all-metal light. (Note that there were also two plastic variants the TL-122B and TL-122C.) The small Buck Rogers special is a hand generated light. By squeezing the lever this would charge the battery thus giving 20-30 seconds of light. Above this is the Traffic lamp used by MP's and sentries. And lastly is another variant of Traffic lamp seen frequently carried on the folding BSA Bicycles.

130

Chapter Nine: Personal Items of the Airborne

Left: Issue cigarette tin. Right: Original "fags" (British slang for cigarette) in their issue tin. These are of the Lister company – many brand names were issued to the troops.

Variants of dressings: the large shell dressing with white shoulder tie which was carried in the rear twin pockets of the Parachutist trousers; the first field dressing carried in the front pocket of the Battledress and parachutist trousers as well as some variants of KD trousers and shorts. And below is one of the dressings carried in the larger ambulance packs.

Here are the 5 finger common issue mittens (gloves) – some have size labels sewn to their inside wrists. Also, most are found sewn by a single strand to keep a pair together until issued.

Left: The clothes brush which was carried in the haversack. Most are stamped with the manufacturer, date and broad arrow.

131

For King and Country: British Airborne Uniforms, Insignia & Equipment in World War II

Left: The full size camouflaged face veil. Many were cut and tied to rifle barrels by snipers. Some were also used as helmet and beret covers. It was somewhat common, especially at Arnhem, to see a trooper wearing his beret with a face veil tied over it for camouflage as the beret on its own would stand out. Right: The rolled up face veil worn as scarf.

Identity Discs

A set of three were issued, two round red discs and one olive octagon disc. One round disc was tied to the light weight respirator. The other two were tied as pictured around the neck. On each disc was the soldier's first initial, last name, religion (example: CE = Church of England), and Army number. Each number would start off with that of his parent regiment (the regiment you were first posted). As a side note every item of kit had to be stamped with the soldier's number. Things had a habit of disappearing in the army if it wasn't nailed down or identified. For instance if a soldier needed a clean pair of socks and he happened to see a stray pair lying about, well he'd pick them up and make good use of them. This wasn't stealing in the soldier's mind, rather it was scrounging, yet it was forbidden by regulation. The discs were made of a fibre like pressed paper. The cording used to tie the discs was like that of the Jack Knife lanyard. Upon death, one disc was left on the body as the other was taken back to HQ and Graves Registration.

Whistle and Lanyard

Issued to each officer as a means of gaining the attention of his troops. It could be used to signal an attack or simply to quite down the men while on parade.

One water bottle

Covered with serge or felt which could be soaked to keep the water cool. Each bottle had a cork stopper with a cord and pin to prevent its loss. This was attached to a small ring at the mouth of the bottle neck. An extra water bottle was often carried in the haversack. Mostly by "old salts" who knew better!

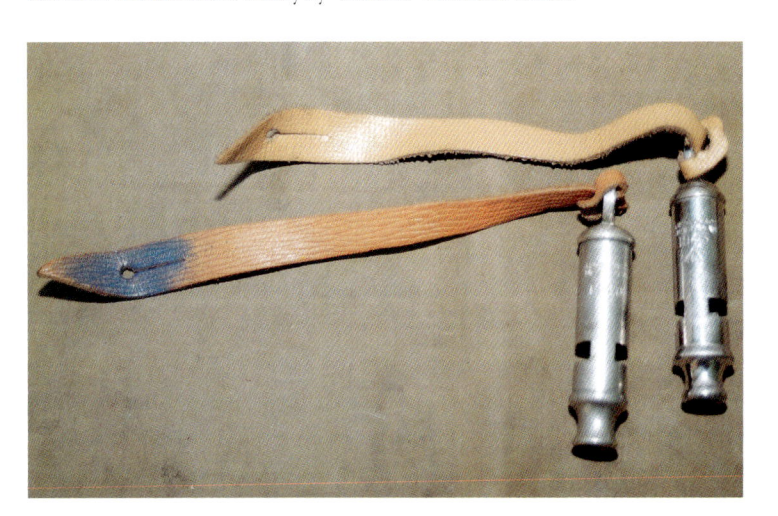

Issue Identity discs. Note how the red disc is worn on its own loop as per regulation. Stamping on the discs: surname and first initial, religion which in this case is Presbyterian, and Army Number which the first series would be of his parent regiment.

Right: The officer's whistle. Stamped into the whistle are the manufacturer, date of manufacture, and broad arrow. The leather piece is a lanyard that fit over the button of the breast pocket or shoulder strap of the officer's battledress.

Left: Shown here is how the leather lanyard attached to the whistle and the button hole on the opposite end that fit over the button of the blouse.

Chapter X

Specialist Airborne Equipment

Life vests

Two patterns were used, the most common was the special MkIII Airborne life vest. The second was a standard infantry life vest, and was the same as those used by the troops wading ashore at Normandy. I have seen photos of the 3rd Air Landing Anti Tank battery wearing the infantry life vest. Glider Infantry wore both variants. Paratroops had to wear the special vests because of the weight of their equipment. The smaller infantry vest would not keep them afloat.

A Captain wearing the experimental body armour. He wears the 3rd pattern helmet, and has his map case to his side as he takes a smoke.

Left: A group of Polish Medics wearing the Airborne Life Vests. Soldiers of the Polish airborne wore their berets pulled, back rather than to the side like the British.

Sten bandoleer

The first pattern bandoleer was manufactured of a lighter weighted webbing than the more commonly seen 2nd pattern bandoleer. The 1st pattern uniquely featured a large single flap that secured the individual magazines and was first issued in 1941. The 2nd pattern was manufactured as early as 1942 of the standard weight P-37 webbing and featured individual closing magazine pockets that were each secured by a tongue and metal loop. Each bandoleer held seven magazines of twenty-eight rounds (9mm) each. Both the British and Canadians made and used them throughout the war.

The full layout of Sten accessories. From the left is the Sten Bayonet frog with bayonet, scabbard and clip on loader, the 2nd pattern bandoleer with Sten magazines, and above is the British box magazine loader, featuring a brass loading handle. Just visible is the stamping underneath the bandoleer's shoulder strap. The smock is a 2nd pattern as noted by the wrist tabs.

The stamping on the Sten bandoleer. Very few are dated 1942, with 1944 the most common date seen by collectors.

A Sten Gunner of the 2nd Parachute Brigade in southern France. The trooper wears the 2nd pattern Sten Bandoleer with its individually secured magazine pockets. Just underneath the right arm is the sleeve water bottle carrier. The smock is first pattern.

Sten bayonet frog

There is a disagreement about this piece. One opinion is that this special frog with added pocket was for the British made clip on magazine loader. The other opinion is that the pocket was designed to carry the spring retaining plate (back plate). This plate kept the spring and rod in place while the butt was attached. Until the butt is replaced the plate prevents the spring from flying out. This is because of the tremendous pressure of the spring. Many Paras jumped with their Stens dismantled and held/secured under their parachute harnesses. I have only seen this frog of British manufacture. Very few were made, with approximately 1,500 pieces manu-

Right: Seen here is the clip-on Sten Magazine loader being removed from the Sten Bayonet Frog pocket.

Chapter Ten: Specialist Airborne Equipment

factured. This was not issued to each man who carried a Sten, and it is a very rare item.

Of special note is that the Hartenstein Museum has in its collection an interesting piece of webbing that was dug up in the surrounding area. This piece is a partial web belt with the Sten bayonet frog, and in the frog pocket is ... the clip on loader, as it was left during the battle of Arnhem.

Sten magazine loader

There are two patterns and three variants. One is the easier to use 'box' loader which fits over the top of the magazine. This was made by the British and Canadians. Another pattern was the British made 'clip on' which some say was to fit into the Sten Bayonet frog pocket. A magazine loader was a must, as after the fifth round it is very difficult to hand load a magazine – there is simply too much pressure from the magazine spring.

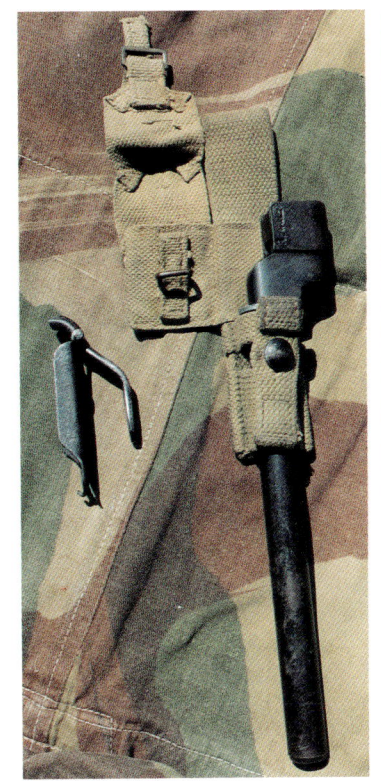

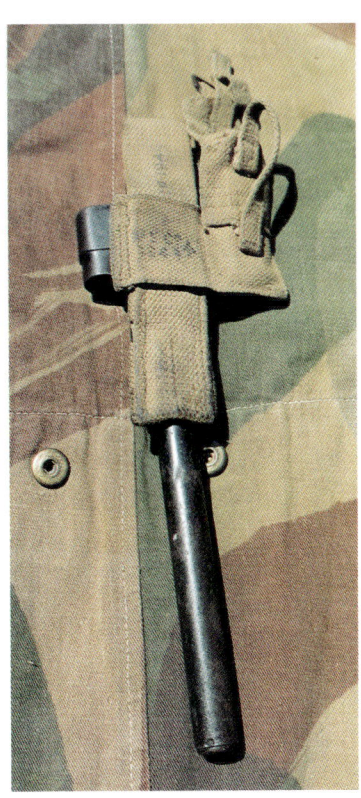

Left: Here is a close up of the Sten Bayonet Frog, bayonet, scabbard and British made clip on magazine loader. Right: The reverse of the Sten Bayonet Frog and its stamping. The thin loop on the right is to secure the pouch to the waist belt and keep it balanced for easy access of the loader, or Sten back plate. The bayonet section also has a loop to fit over the belt. Note the stamping M.E.C. manufacturer, 1944, broad arrow.

Seen here is how both loaders attach to the Sten magazines. Each loader has a loading handle which sends the round into the magazine and then comes back up to drop in and send down the next round.

Just visible is the stamping on the box loader, and a broad arrow also stamped into the brass handle. The clip on the back of the loaders are for removing the loader – pull back or push up and the loader comes right off.

Leg bag

First issued in 1944, the Leg Bag was designed to carry the parachutist's equipment and small arms, thus allowing his freedom of movement while in descent. Special 'quick release' straps secured the bag to his leg which was further secured by a gap for the soldiers boot to fit between on the bottom of the bag. Out the door he'd go with the bag attached to his leg and resting on his boot. As ground rush began, the bag would be released by pulling a special cord. It would then dangle below the Para until it hit the earth below. Upon landing the Para would open his bag, assemble his kit and off he'd go leaving the bag cast aside along with the oversmock. The bag was heavily padded on the bottom and was also used to carry most anything the soldier desired that was not to be carried in a pannier and/or drop container. These were also issued to the American Airborne for Normandy. The bag measured 30" in length and 14 1/2" in diameter. It was open on one end and down one side the entire aperture was fitted with brass grommets and laced with cord. The inner cushion pad was about 4" thick.

One method of carrying the box loader underneath the Sten Bandoleer shoulder strap.

The Leg Bag, which is one of the hardest pieces for Airborne collectors to find. To the front are the web leg straps, and to the right the quick release cord that, when pulled, releases the bag and enables the bag to hang below the trooper during descent.

Above: Just visible is the white tape which is part of the quick release cord.

Left: Note the 1944 date on the inside of the Leg Bag. 1944 dated kit is highly prized by collectors.

Chapter Ten: Specialist Airborne Equipment

Enfield Valise
Made of thick padded felt, it was designed to work the same as the leg bag strapped to the leg and body and then released to hang 20 feet below the descending paratrooper.

Right: As the Enfield rifle could not be broken down and carried like the American M-1 Rifle and it's airborne Griswald transport bag, a special padded valise was designed. Here is the front view of the bag. To the left is the Quick Release handle which is made of khaki webbing. In the middle is the "paying out rope" and rope pocket. Once the trooper pulled the quick release with his left hand, he would then pay-put (pull out) the rope with his right hand until the valise was suspending 20 feet below him. Note the small web strap and snap-seen at the top right end which helps secure the weapon in the valise. (Courtesy ABM Hartenstein).

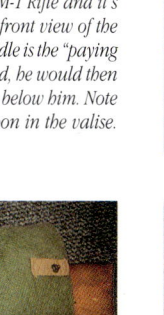

Left: Underneath the valise is a web loop which attaches to the parachute web harness-thus securing the valise to the trooper once the paying out rope reaches it's end (20 feet below the man). Right: Here is the reverse of the valise. Note the securing strap at the far right. The bottom opening folds over and the strap is secured via it's snap, thus containing the rifle. (Courtesy ABM Hartenstein).

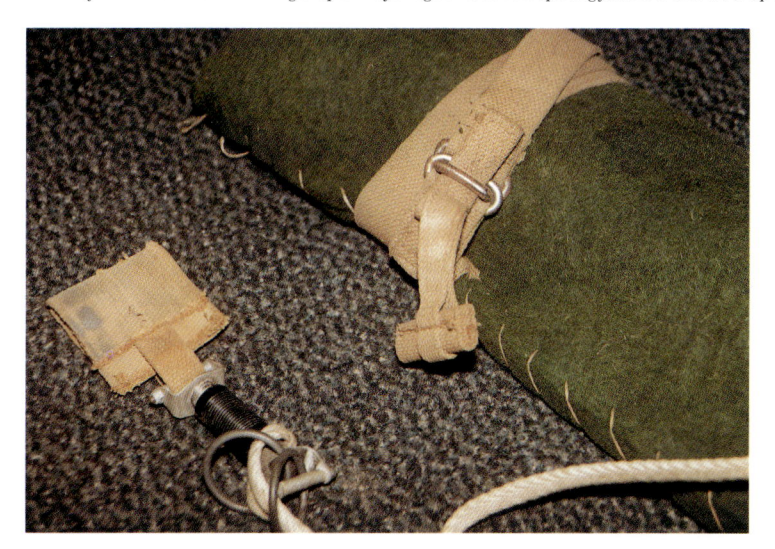

Left: A close up of the harness loop, quick release handle, and shoulder/neck strap which was actually worn around the mans neck/shoulders to further secure the valise while jumping from the aircraft. As can be seen the valise is simply a piece of thick padded felt folded over and its edges sewn together. Right: Another look at the harness loop. (Courtesy ABM Hartenstein).

Left: The rope pocket. Right: A closer look at the securing snap and how the paying out rope is secured to the valise-thus preventing it from ripping away from the valise while in decent. (Courtesy ABM Hartenstein).

137

Bren Valise

Of the same idea as the Enfield valise, yet larger in size to accommodate the weapon.

Right: Here is the LMG (Bren) Valise. Designed a bit differently from the Enfield valise, the trooper would pull the quick release with his right hand and would then pay out the rope until the valise was suspended 20 feet below the Para. The quick release handle is the same as found on the Leg Bag. Below: Reverse of the LMG valise. (Courtesy ABM Hartenstein).

MKII Chest Respirator

Specially converted for the Airborne to be worn upside down so that the wearer could simply open the bag and the mask would fall out, enabling the wearer to put on his mask faster should he be subjected to gas upon landing by parachute. There was a special loop and snap added to the bag for this purpose. This is probably one of the rarest items to find. Also carried in the bag were anti gas eye shields, anti dimming ointment and anti-gas ointment. Small pieces of cotton were used to clean the lenses.

Light weight respirator MKIII

It was carried in a waterproof bag which was attached to the waist belt. In the bag were anti-gas eye shields, anti-dimming ointment and anti-gas ointment. Small pieces of cotton were used to clean the lenses. Note that there were anti-gas brassards that were issued to the 2nd Parachute Brigade for the Southern France operation. These brassards were specially treated to react to gas by changing color, thus alarming the wearer of its presence.

The quick release (handle) and rope pocket. The web strap is worn around the neck/shoulder as the bottom strap is secured around the right leg. As the LMG weighed quite a bit more than the rifle, one needed the extra support while jumping from the aircraft (and until the bag was released). (Courtesy ABM Hartenstein).

The bottom of the LMG valise and the lower quick release pin. Once the handle was pulled both the upper and lower (as seen here) pins would release the web straps-thus the valise would fall from the trooper and suspend underneath him. This is basically the same manner in which the Leg Bag was released. Note that rather than a strap and snap the LMG is secured via a web strap and metal buckle. (Courtesy ABM Hartenstein).

Left: The MK III light weight respirator bag as worn on the back and attached to the P-37 waist belt. The loop on the right side of the bag is for a shoulder strap (yet the bag was worn this way by regulation), attached to the belt via the brass hooks. Note the very faded 1st pattern Denison.

Chapter Ten: Specialist Airborne Equipment

Above left: Contents of the little green bag: ointment for the skin, lens cleaners, protective goggles which are worn underneath the mask, and the actual gas mask. These masks do not fair well with age as the rubber cracks and can become brittle. It is very difficult to find a mask in mint condition, especially the MK III. Some troops were known to throw these masks away and stuff personal items into the bag.

Above right: Seen here are the skin ointments and the small lens cleaning cloths.

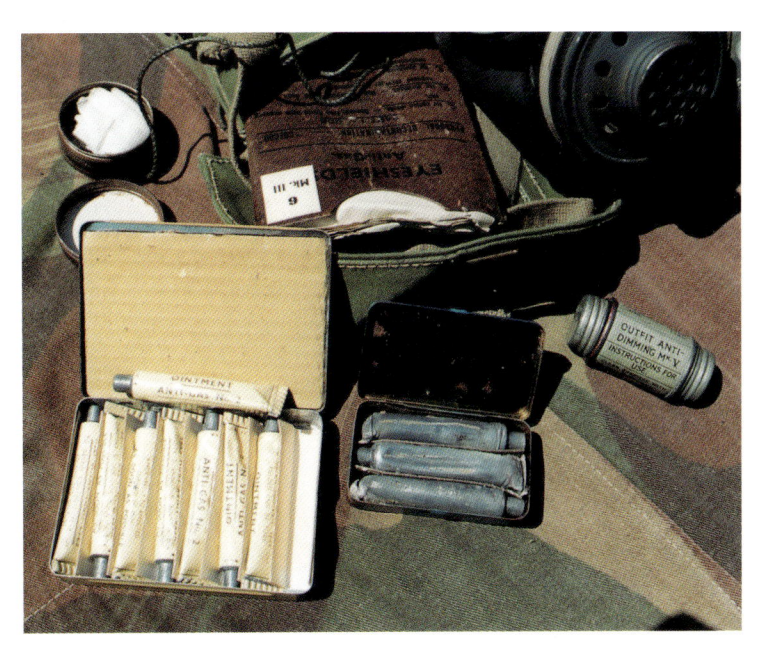

Above: The MK III Respirator bag. Note the shoulder strap, when not in use it is kept rolled up and inside the bag. The stamping on the respirator bag is found as you see here, on the inside cover.

Folding Trolleys

Designed to be a carry-all for ordnance, equipment, and supplies. There were several variations, the author having seen up to three.

A folding Airborne trolley. Note the toggle ropes, which were tied to each front corner. The trolley was pulled by a team of men as they were at times 'loaded' with kit. These ropes are the thinner variant. Also seen is where each part of the frame fits into each other and screws into place. Also of interest is the Para in shirt sleeve order and his typical large war time beret, as the post-war are a dead give away being 1" to 1 1/2" shorter in the crown, and come to rest above the ear.

139

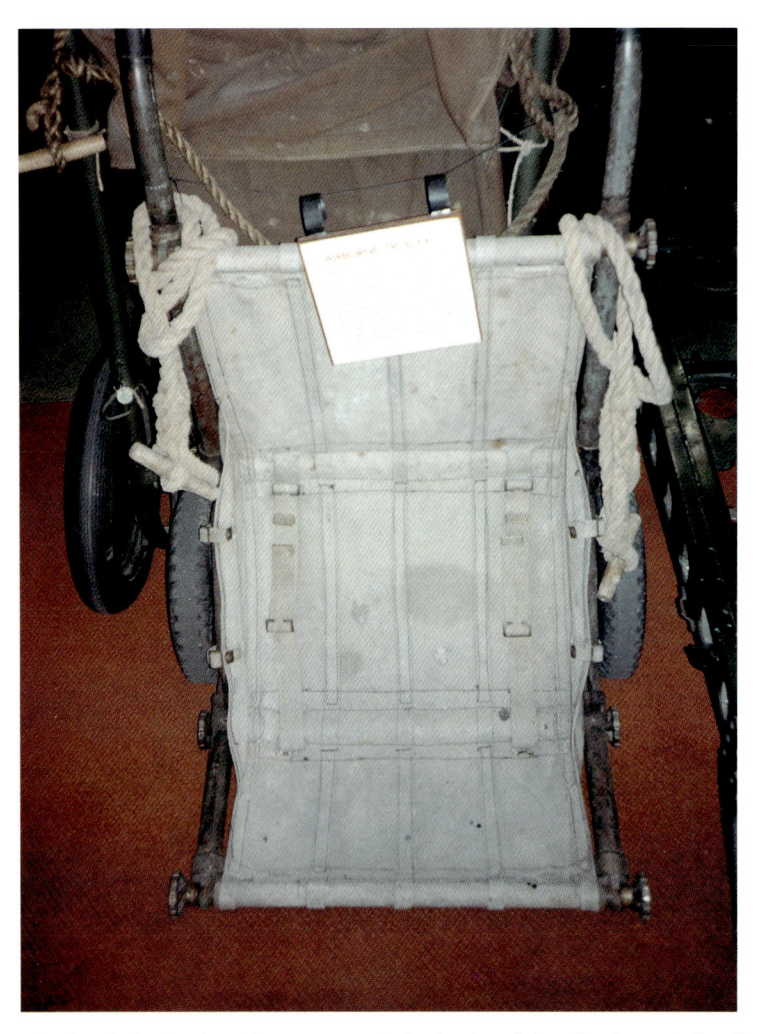

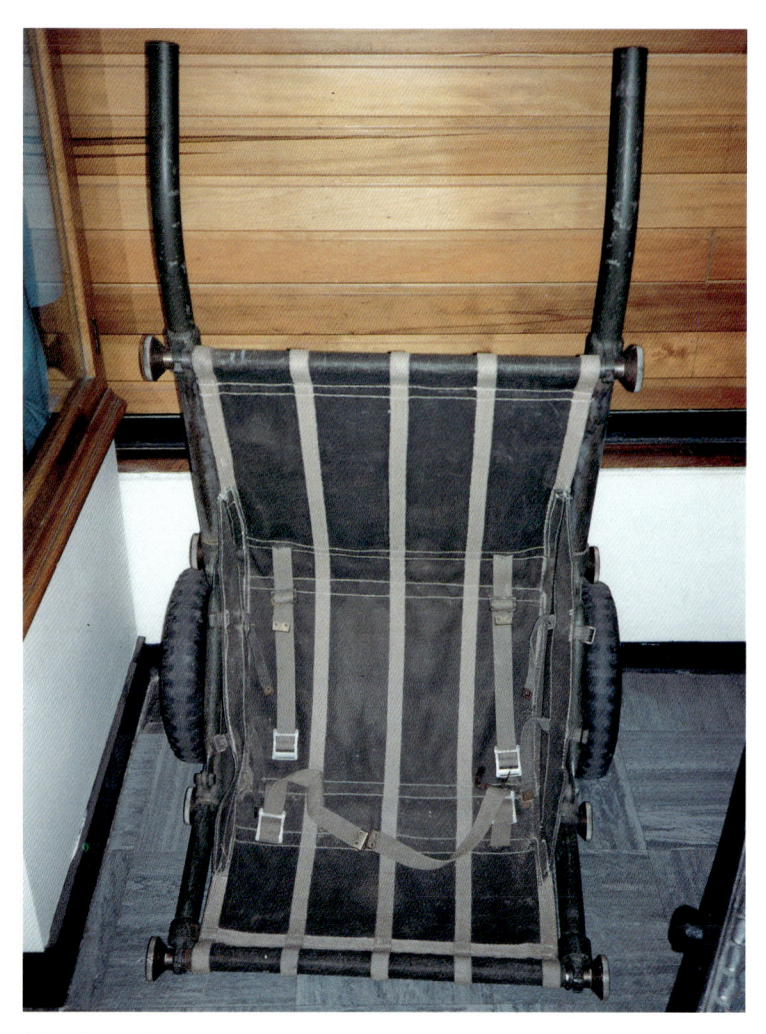

For King and Country: British Airborne Uniforms, Insignia & Equipment in World War II

Left: A better look at how the toggle ropes were attached and used to pull the trolley. The rounded handles hold the trolley together and also enable it to be dismantled. Right: An excellent example of a trolley. Again note the simplicity of construction and assembly.

Body Armor

Issued as an 'available extra' for the battle of Arnhem. It was worn by both the Poles, Glider Pilots and Recce Squadron Drivers, and was usually worn underneath the Denison and over the BD.

Right: A set of armour that has not been blancoed and in its original color. Above: The reverse of the armour. Note the padding for both comfort and impact absorption.

Chapter Ten: Specialist Airborne Equipment

Above left: This officer's Batman must never sleep as this set of webbing and armour is immaculately blancoed and polished.

Above right: A closer look at the buckles and hooks of the armour set. Also note the shortened tanker's holster – this was by regulation, the cutting off of the long 'gunfighter' sling strap, and folding over a bit to make a loop for which the waist belt fit through. Note the cleaning patch slit in the revolver cleaning rod.

Right: Another look at the rear plate of the armor set. This portion was to protect the sides, as the front protected the chest and groin.

Bergen Rucksack

Widely used by the glider pilots, as they were not issued with the general P-37 webbing 'rig'. A Glider Pilot would attach the Bren auxiliary pouches to the pack and he was off. Bicycle troops of the Airlanding Brigade and Commandos also used the Bergen.

Left: Seen here is a 1944 rubberized Bergen as worn by the Commando Assistant Bren Gunner. Of interest is how the web straps are coiled and secured on the rucksack. Around his shoulder is the Bren spare barrel bag which hold an extra barrel and, in his right hand is the all too heavy metal Bren magazine chest. On his back the Bren Gunner has the earlier variant Bergen of 1942 which was khaki and non-rubberized. The Bren gunner wears the Denim overalls, over his battledress. The A-gunner also wears the overall trousers.

141

For King and Country: British Airborne Uniforms, Insignia & Equipment in World War II

Left: The Bergens side by side. On the right the earlier non-rubberized of 1942, and at left the olive rubberized of 1944. They could, and were, filled to capacity by those who used them. Not only was the soldier to carry his personal items and provisions, but also other equipment needed in the field – veterans knew that resupply could take longer than expected. Also of interest is the Bren Gunner's BD seen underneath the denim overall blouse. Right: The full back of the 1944 olive rubberized Bergen can be seen here. Note how the toggle rope was attached to the rucksack, commonly done in this fashion. Also of interest is the hessain sacking used as camouflage underneath the helmet net.

Toggle rope

Issued to each man and either worn around the waist or over the shoulders and secured in the rear. Some ropes are were hung from the back of the Bergen Rucksack or Haversack. The idea was that a body of men could form up and make a rope ladder or rope bridge, or use the ropes to aid in getting a jeep out of a ditch. There are two variants, a thinner 1/2" hemp rope with small 4" to 5" wooden toggle. These are mostly seen in use by Commandos. A second variant, which was the more commonly seen, had a 1" thick hemp rope with a larger 6" wooden toggle. Some wartime ropes will be either war dated and or broad arrow marked on their toggle. Other variants are possible yet these are so easily made that one could make one in school so be careful of fakes. The actual wooden toggles are the telltale sign as these are difficult to reproduce properly.

Above: Polish Commandos using toggle ropes. Both are armed with the No.1 MK III rifle. The Commando lending a hand is wearing a leather jerkin which served as a cold weather coat for such specialists. Just visible are his shoulder titles above his Co Ops badge.
Right: The toggle rope. Though the toggle does look correct, this may be a reproduction due to the fraying of the rope.

142

Chapter Ten: Specialist Airborne Equipment

Right: Polish Commandos of No.10 IA using toggle ropes to scale a cliff. Note how each toggle is fitted through the loop of the next thus forming one long rope. Of interest, on the right shoulder of the climbing Commandos blouse, is the Combined Operations badge in its as issued 'tombstone' shape.

Folding Bicycle Pack Frame

Above: The folding bicycle pack frame. 1944 is the earliest dated that the author has seen. The frame is commonly seen on the front of the folding bicycle, which was commonly used by troops in Normandy. Note the manner in which the leather straps are coiled and secured. The Bergen Rucksack fits on top of the canvas shelf, and the leather straps were done up over it. This could also be worn on a man's back as a manpack to carry heavy ammunition boxes for instance.

Right: The stamping: manufacturer, date and a very faded broad arrow. Note how the frame is cured and fits the Bergen perfectly.

143

Flight Crew Thermos and Thermos Carrier

Airborne sleeping bag

Of the same pattern as the Denison, yet printed on heavier denim material. Note that fake Denisons have been made out of sleeping bag material and made to pass as genuine smocks. There is a distinct difference between the twill of the smocks and the denim of the bags. The sleeping bag was designed to keep the Paratroopers warm while in transit in the aircraft, enroute to the jump. A Para would sit inside the bag with his entire kit and chute on, as the sleeping bag has a wide opening and opens to the bottom with a series of toggles and twine loops. The earliest bag I have owned was dated 1942. There is little information about these bags. They were not taken into combat by the paratroops, though they were taken in by glider, mainly by the RAMC for the wounded. Each bag is lined on the bottom with a water resistant pad. The date stamp is found stamped on this bottom lining, and is usually a simple year date and broad arrow.

Above: This is the Airborne/Flight Crew Thermos. In it was carried hot soup or tea for the crew and troops to sip while in flight. Note the brass snap which secured the carrier's lid to its body. Also note that most of these thermos labels have not faired well with age. Below: The padded carrier for the Thermos. This was to protect it, as the flight could be bumpy and/or hectic from bad weather or enemy flak. (Courtesy of the DeTrez collection)

The Airborne Sleeping bag, dated 1942. Note the water proofed lined bottom. The date stamp is very simple and is placed on the bottom end. Of interest is how it is laced up the front with a series of wooden toggles, string and brass eyelet. It was designed so that the Paratrooper could sit inside it while in flight with all of his equipment comfortably. Though they did not drop into battle with the trooper, they did arrive later by glider with the Medical and Divisional troops. Note that it is very similar to the Denison pattern, yet is printed on a heavy denim rather on twill like the Denison Smock.

Chapter XI

Airborne Vehicles

Vehicles will be covered in greater depth in a later volume. The following however are most relevant to the airborne units.

Wellbike

A collapsible two stroke motorscooter, perfect for reconnaissance and messenger duties. It was simple to fold up which was done by withdrawing a pin prior to pushing the saddle down until it rested on top of its pillar tube. The handlebars could be folded back as the steering head collapsed. From here it could be placed in a container for an operation. It weighed about 70 pounds. Eleven seconds was the official time needed to unfold the bike and start the motor. Its top speed was 30 miles per hour. The bike was first issued and put into service in late 1943 and saw use throughout the war.

The Welbike inside an open drop container. Frames and forks were designed to withstand the severe shock of landing. The crash pan was the first to hit the ground. This pan was a semi-metal bulb attached to the bottom end of the container (keep in mind that when dropped by parachute this container falls vertically). The bike was stored with its rear wheel at the crash pan end. The parachute was packed and contained at the open end and secured by the metal D rings seen here.

General Sikorski inspecting members of the Polish Brigade. This was a Reconnaissance platoon as all are equipped with Welbikes. Note the Sten bandoleers carried by each cyclist. Notice the Canadian gas capes attached to the top of the haversacks – as ground sheets are not visible here, the gas capes must have been issued in place of them.

Folding BSA Bicycle

This bycycle was widely used, by not only the Airborne, but also the Commando and line Infantry regiments. Able to fold in half it was carried and 'jumped' with troopers into battle. A special line was attached to the bike similar to the leg bag. The bike would dangle below thus freeing the paratrooper's arms and hands during descent. A rack fitted to the front of the bike carried the Bergen Rucksack, and a rack built onto the frame held the soldier's rifle.

A member of the Recce Squadron with a folding BSA. Note his smock has added hose tops at its cuffs, as this is a 2nd pattern with wrist tabs. The BSA tool kit hangs underneath the seat. Also of interest is the folded ground sheet held, as by regulation, in his haversack.

The BSA folding bicycle, weighed 26 1/2 pounds and was used throughout the war, not only by the airborne, but also the Commando and selected line units. The folding BSA could be dropped via a special 'Q' parachute, though more commonly the bike was carried in the gliders, and/or dropped with the paratrooper. As mentioned earlier, if the bike was jumped, it was done so like the leg bag and released via a suspension cord prior to landing.

Right: A wounded Glider Pilot pauses for a short rest. Of interest is the folding rack attached to the bike of which is a 3" mortar bomb carrier. A small tool kit hangs underneath the seat.

Chapter Eleven: Airborne Vehicles

125 Royal Enfield Flying Flea

Light weight motor bike used by dispatch riders (messengers) and reconnaissance troops.

Above: The MK I dispatch steel helmet with riders goggles, as worn by the Airborne.

Right: A Polish dispatch rider of No.10 IA Commando. Of interest are his cyclist gloves and first pattern 'pulp' dispatch helmet – similar to those worn by the Airborne dispatch riders. This was the first pattern dispatch helmet, and was made of a hardened paper, and was about 3/4" thick. It was replaced in July, 1943 with the second pattern MK I dispatch helmet which was made of steel and had a thick leather padded liner.

The Royal Enfield Flying Flea in airborne colors. It was delivered both by glider or parachute in a metal crash proof frame with parachute attached. The Flying Flea's particulars are as follows: single cylinder 2 stroke, three speed (22.7 to 1, 12.7 to 1, and 7.8 to 1), 4500 r.p.m., a single dry plate clutch with cork inserts, hand operated external expanding breaks, 2.77 x 19 tires (2.50 x 19 early type), wheelbase 4', ground clearance of 4 1/2", and a tank capacity of 1 1/2 gallons with a radius of 140 miles. The bike was 6'2" long, 1'4" wide (folded), and 2' 10 1/2" tall. It weighed 125 pounds dry (without fuel).

Bren Carrier (Universal Carrier)

The Bren Carrier was used in a variety of roles from towing the Six Pounder Anti-Tank Gun, carrying supplies and munitions or troops. The Bren Carrier could only be carried by the Hamilcar glider.

Right: A Bren carrier at Arnhem. To the front is a Bren and padded airborne thermos container, and two spare carrier wheels. To the soldier's right is a BSA, and to the right of that on its side an Enfield Flying Flea. Note that the carrier is loaded to capacity with 'all sorts'. Each Airlanding battalion had four carriers, and were transported via Hamilcar glider, two carriers per Hamilcar.

Jeep, Airborne

Lend-Lease (on loan, and or in trade for the use of land in the UK for U.S. bases) from the U.S., the jeep saw wide use by the Airborne as an all-purpose, hold-all like the Bren Carrier. It was used in carrying troops, supplies and towing the six-pounder anti-tank guns and the 75MM Pak Howitzer. The British manufactured their own jeep trailers which were deeper and larger in size than the U.S. trailers. This enabled more to be carried within them. A special waterproofed cover was also made for this trailer. As these jeep were brought in by glider, any 'extras' had to be removed to help lighten the load while in flight. This was done by removing rear view mirrors, cutting off the bumper guards and removing the windscreen. Most of the Airborne were equipped with these jeeps. Examples being the RAMC who added special stretcher racks, the Recce Squadron who heavily converted the jeep including the mounting of Vickers 'K' machine guns and the RASC, who did not make 'special' conversions yet were very dependent on the jeep for the collection of supplies.

A typical Airborne jeep as would be found on any given operation. Note how the windscreen has been removed and the front bumper shortened. On the jeeps bonnet (hood) are the metal Bren magazine chests, and a spare Bren barrel and carrier. Notice that the Jerricans are held behind the seats – one of many special Airborne conversions.

Of interest are the Vickers drum magazines which were fed to the Vickers 'K' guns. Just visible is the bonnet frame which is attached to the bonnet, and also the cut front bumper.

Seen here is a bonnet mounted frame in which ammunition boxes could be stored. A later volume will go into greater detail on the airborne jeeps and other vehicles.

Chapter XII

Airborne Weapons

From the top left: the No.4 bayonet, below that the Sten M.C. bayonet MK I, at the right is the MK V Sten with wooden stock and foregrip, below is the MK II Sten, the No.4 rifle, the No.4 MK I (T) rifle, and lastly the Bren LMG MK I with its bipod in the up position. (All Courtesy David Gordon collection)

For King and Country: British Airborne Uniforms, Insignia & Equipment in World War II

Bayonets and Daggers

Bayonet No.1 and No.3
Both are referred to as sword bayonets as their blades are 17" in length. These accompanied the No.1 MK III Rifle.

Bayonet No.4
Manufactured to accompany the No.4 MK I Rifle and Mk V Sten. Its blade was 8" in length and 10" in its scabbard. There were four variants: MK I with cruciform section blade; MK II with round spike and one piece blade; the MK II* of two piece construction; and the MK III with a rougher fabricated and welded socket.

Bayonet No.4 Metal Frogs
There are four variants: Mk I with tapered steel and ball end; MK II with parallel sided sheet steel with a hole in the tip; MK III with a plastic parallel sides body with an alloy mouth piece; and lastly the MK V which was American made and featured a plastic tapered body.

Bayonet No.7
Designed for both the No.4 Rifle and the MK V Sten it's blade was 8" and with it's scabbard 12/3 inches. This is an unusual piece as it features a plastic hand grip. This bayonet was introduced in early 1945 in time for the Rhine Crossing.

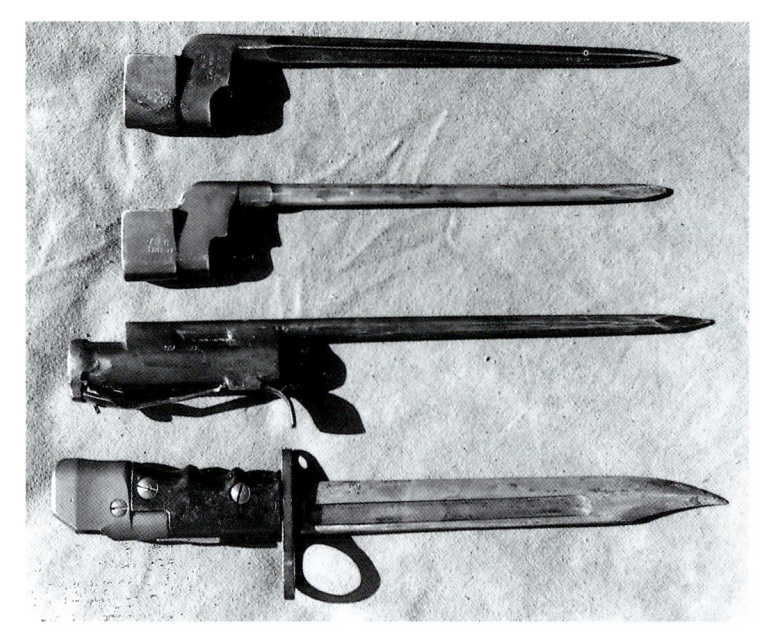

The No.4 and Sten bayonets, from the top down: No.4 MK I Cruciform, No.4 MK II, Sten M.C.MK I, and the No.7.

Bayonet, Sten M.C. MK I
Made specially for the MK II Sten it's blade was 8" and with its scabbard it was 12". This is a very rare piece as few were manufactured. Introduced in 1942 it saw little use when compared to the No.4 Bayonets.

Faifbairn-Sykes Dagger (fighting knife)
Commonly known as the F.S. dagger, this knife was a highly prized item by those who carried it. The knife is comprised of two main parts, the blade, and the hilt which features the tang nut at the end of the hilt, the grip and the guard. Commonly issued versions to the Airborne were the 2nd and 3rd pattern knives. Most that appear in collections, and are actually documented as being knives which belonged

Below left: A 2nd pattern knife for which the nickel plating has been polished off and the brass remains. Note the leather tabs of the sheath, that were to be sewn to the trousers. The nickel plating has also been polished off the sheath's protective tip. The trouser conversion is that of a French Commando and not a Paratrooper.
Below center: Another French conversion to Canadian battledress trousers. This knife is a 3rd pattern. Note how the hilt and sheath tip are blackened – this is similar to bluing a gun for protection against corrosion.
Below right: A 3rd pattern knife in a pair of 1942-44 (as the pocket is chamois lined) Parachutist trousers. Note how the plastic button secures the leather sheath. The elastic band that held the knife in place simply did not fair well and was worn off with wear. The very end tip to the hilt is called the "Tang and Tang nut", and was set in place with a special wrench.

150

Chapter Twelve: Airborne Weapons

to members of the Airborne, are 2nd and 3rd pattern knives. The 2nd pattern knife featured a nickel plated hilt – both the hilt and blade were blackened. Those with nickel plated hilts that are marked 56 broad arrow, and 60 broad arrow were part of a rush order from Wilkinson Sword Ltd. just prior to the Bruneval raid. Blackened hilts are usually marked B2 which was a Wilkinson stamping code. Also the tang nuts feature the special wrench marks of Wilkinson Sword Ltd. Private purchase 2nd and 3rd pattern knives feature both the Wilkinson logo and the FS (fighting knife) logo, one on each side of the upper blade.

All knives were blackened, though this blackened finish wore off from the soldiers polishing them as they like to do. The dagger was either carried in its scabbard and worn on the hip if the wearer did not have a pair of Parachutists trousers, or in the Parachutist's Trouser special dagger pocket.

1st pattern nickel 'S' hilts were mainly issued to the Commando, and by the time the Paras came to be most all 1st patterns had been issued out. No additional 1st patterns were made after this. Those Paras most likely to possess one might be from No.2 Commando as they had made up the cadre of 1st Parachute Battalion.

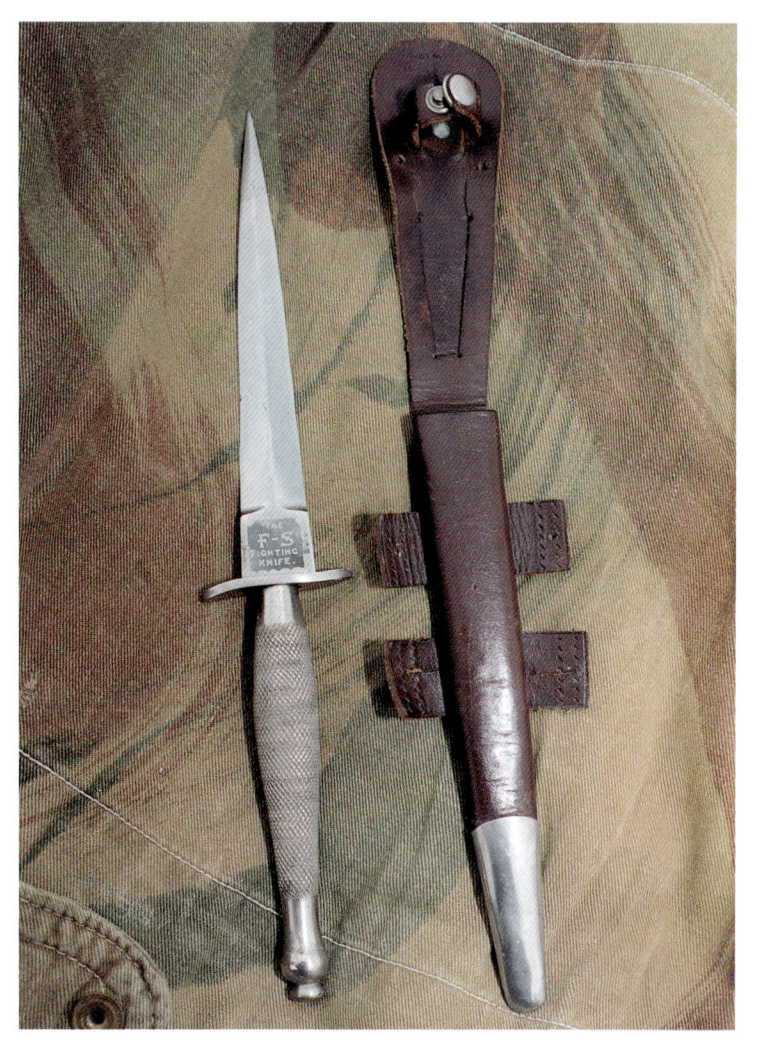

A documented 1st pattern knife that belonged to Bren Gunner George Jones of No.4 Commando, who carried this knife throughout the war. Note the F-S Fighting Knife stamping. Also note the 'S' guard, which is the tell tale sign of a 1st pattern knife. As can be seen the blackening has long since worn off from use and polishing. Of interest is, at the top of the sheath, the leather retaining strap and snap which was not featured on other pattern sheaths.

The reverse of Jones's knife. Note the stamping Wilkinson Sword Limited, London. Note how the sheath is sewn together.

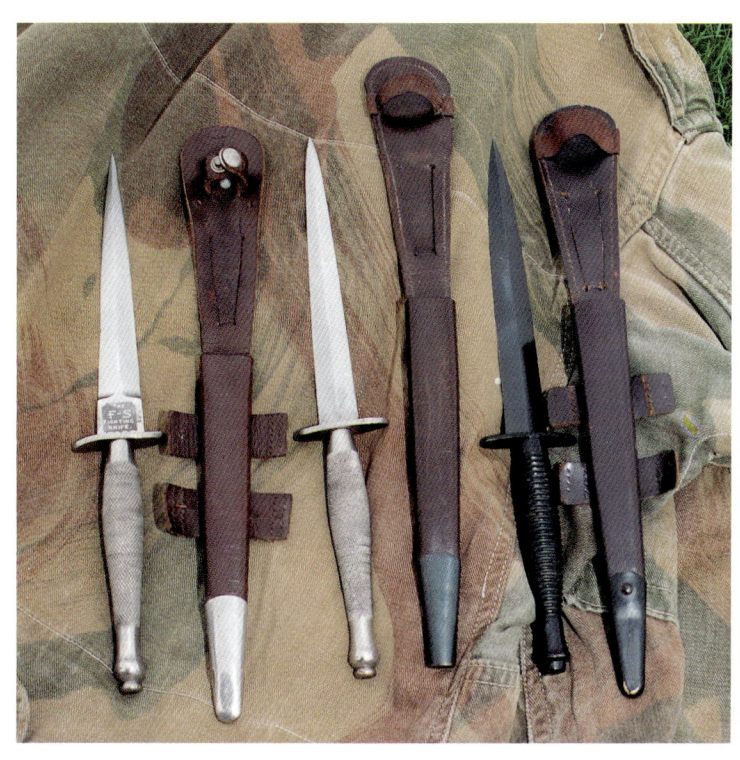

Right: From the left: 1st pattern nickel plated with 'S' guard, 2nd pattern nickel plated, and 3rd pattern which still retains its blackened finishing.

Revolvers and Automatics

Pistol Revolver No.2
Of single and double action this .38 caliber was the general service revolver issued to Officers and Specialists: Bren gunners, radio operators, messengers, and drivers. It had a six round cylinder and weighed 1.7 pounds.

Pistol Revolver, .38
This Mk IV supplemented the supply of the No.2's. The MK IV was of single and double action with a six round cylinder and weighed 1.6 pounds.

.32 Colt Automatic
A Lend-Lease weapon, this smaller automatic was in wide use prior to the issue of the Inglis Hi Power. It featured a 7 round magazine, and weighed 2.03 pounds. General Urquhart carried one at Arnhem, and shot a German who happened to peer through the wrong window while the General was separated from Division and presumed dead or captured.

No.1 MK I Ingils-Browning Automatic 9mm
A preferred automatic weapon of Canadian manufacture and commonly issued to officers. Its caliber was 9mm and weighed 2.3 pounds. Three 13 round magazines would be carried – one in the gun and two in the P-37 Hipower magazine pouch. The Ingils Hipower did not see use until the Fall of 1944. Its range was effective out to 50 to 75 yards.

The Englis Hi Power with lanyard which is attached via a metal ring to the butt of the grip. Note the wooden stock, which was how the weapon was originally issued, yet these stocks were not issued to the Airborne or Commando. This weapon was originally made for the Chinese, and as the contract was canceled the guns were not delivered. Also of interest is the holster hook which attached to the stock, then the web loop fit over the waist belt. Below are two variants of holsters for the Hi Power. The magazines each hold thirteen rounds, which at that time of the war was revolutionary.

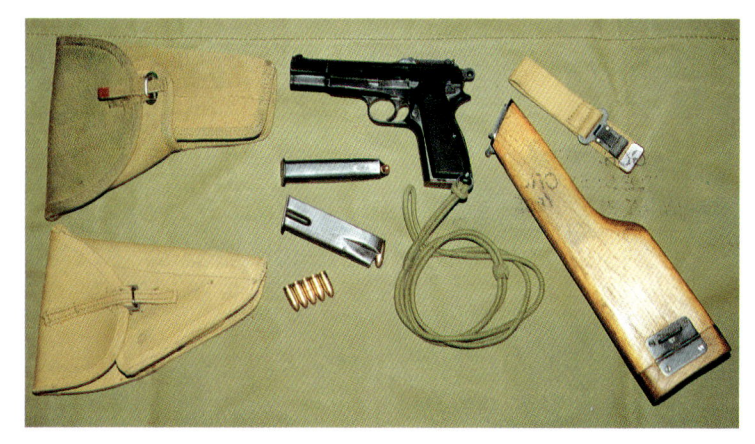

The detached stock and web belt hook.

Revolver No.2 and its P-37 holster.

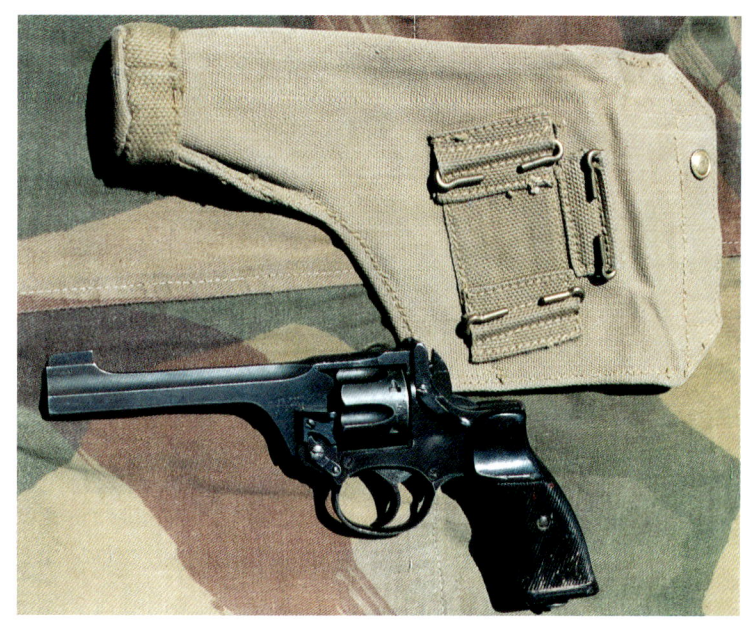

Note the brass hooks on the reverse of the holster. The waist belt attaches to the pair of hooks on each side. Above that is a pair that is for the attachment of the pistol ammo pouch.

1911 .45 Automatic
Manufactured by the U.S. this was a favorite and standard issue to the Airborne. This weapon was part of the Lend-Lease Program. It featured a 7 round magazine and weighed 2.4 pounds. Its range was effective out to 50 yards. The .45 could be carried in the specially lengthened P-37 revolver holster which was 1" longer than the rounded flap P-37 revolver holster.

Rifles

No.1 MK III Rifle .303
Introduced in 1903, this general service weapon of the British and Commonwealth forces of the First World War was kept in service throughout the Second World War. It was the first bolt action to hold ten rounds making it superior to the German Mauser's five round carbine. The British Airborne discontinued using them prior to their going into North Africa on November 12, 1942. This rifle was easily prone to foul when dirty – later models were of a better design and did not suffer this fate. Production of this rifle was terminated in November 1943 as the No.4 MK I had become the general service weapon for His Majesty's Forces. The No1 MK III weighed 8.6 pounds. Effective range was 800 to 1,000 yards.

Chapter Twelve: Airborne Weapons

No.4 MK I Rifle .303

This rifle was first produced in June of 1941 and saw initial action in North Africa with members of the Airborne and British Second Army (as the Eighth Army were the old salts who still carried their No.1 MK IIIs). Used throughout the war (after 1943), this was the best made and quickest firing bolt action rifle of its time. A soldier could pride himself in both firing and reloading all ten rounds within one minute. The No.4 MK I weighed 9 pounds. Effective range was 800 to 1,000 yards.

No.4 MK I (T) Sniper Rifle .303

No.4 MK Is were converted to Sniper rifles by the addition of a No.32 Scope and special butt and cheek rests. Each Battalion had their own sniper teams which consisted of two men per team – one marksman and one spotter who assisted the rifleman in acquiring his target. The need for snipers became especially apparent after encountering the deadly German snipers who lurked in the hedgerows of Normandy. As in Normandy where the fighting at times became static, it was a necessity to fight fire with fire. The No.4 MK I (T) weighed 11.6 pounds.

The No.1 rifle with its stub nose – 8" from the tip of the barrel is the lug where the sword bayonet attaches.

A trooper loads his No.4 rifle. Note how he pushes down with his thumb the five round charger. Around his neck are two bandoleers of .303, totaling 100 rounds. Of interest is that only one rank chevron is worn on his Denison, which was very common among the Airborne.

Note the latch in the butt of the stock, as this is where the pull through and oiler are kept. The trooper is now removing the pull through (the rifle cleaning kit), which consists of a brass dowel attached to a length of cord. The brass end is dropped down the barrel and a small patch cloth is attached to the other end of the cord. The weight of the brass dowel pulls the cord through the barrel, thus the patch cloth absorbs any dirt in the barrel.

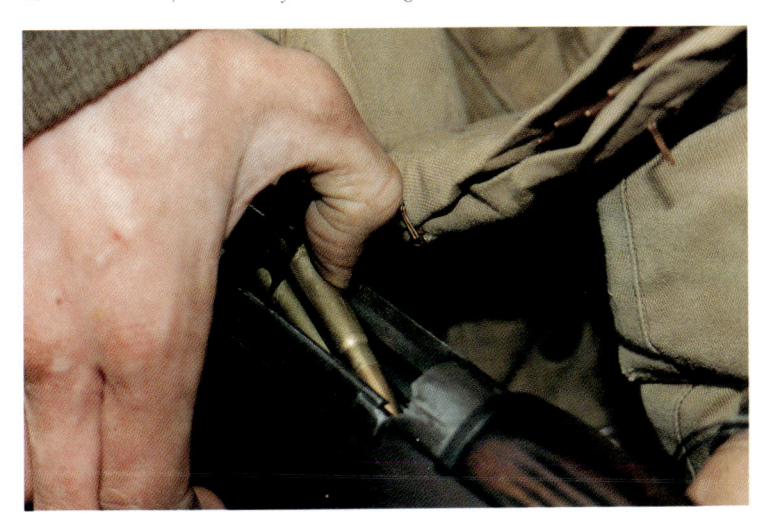

Left: The trooper pushes the round down and from the charger, as once the rounds leave the charger, the charger is removed and kept in a pocket for reloading it later. Note the ball ammunition in the bandoleer. Also interesting on this 1st pattern smock is the short wool cuff – this was as issued!

Here the brass oiler is removed from the stock.

The oiler is unscrewed and dipped into the brass tube – oil is then applied to the weapon.

The Sniper was a trained marksman who possessed a high degree of Field Craft. This is the knowledge and art of maneuver and terrain. A Sniper had to be observant and able to pick out the best targets as well as knowing when not to shoot but rather waiting for a more advantageous opportunity. He had to make precise and singular shots as each one had to count. He also had to be a good map reader and have a good knowledge of enemy weapons and identifications. A good Sniper was able to hit a man's head from up to 200 yards and his trunk (chest/upper back) body from up to 400 yards. In good weather he may be able to hit a man's body from up to 1,000 yards though this was discouraged unless specifically necessary. Officer's and NCO's were the prime targets as the Germans were not taught to act on their own but taught rather to rely upon their leader's orders.

A Sniper was usually outfitted with: one No.4 MK 1 (T) Rifle, one No.32 rifle scope and case, one telescope with carry case, two face veils (one to be worn and a second to drape over the rifle), a liquid prismatic compass, a general service watch, prismatic binoculars and case, adjusting tools for the rifle which were the No.1 MK 1 and the No.2 MK 1, and lastly the Denison smock. If working as a team and not simply the Sniper, the Spotter would be equipped with: 50 rounds of S.A.A. (.303 small arms ammunition), 2 grenades being either No.36 or No.77, 5 rounds of tracer, 5 rounds of A.P. (armour piercing), one water bottle, and one emergency ration.

As the perimeter closes in sniping has become rampant on both sides, the goal of causing as much panic and trouble as possible until the final push is made to wipe out the Airborne. Sniping in Oosterbeek was at an all time high towards the last few days. The Airborne snipers did their best to fight fire with fire making the enemy pay for every inch of ground.

Left: The sniper takes his time and waits for the shot.

Chapter Twelve: Airborne Weapons

S.A.A.

Small Arms Ammunition was that type used for the pistol, submachine gun, rifle, LMG, and MG. S.A.A. rounds were complete as they included the cartridge case, percussion cap, propellant charge, and the bullet. Pistol calibers such as 9mm, .38, and .45 usually shot ball. The rifle, LMG and MG shot .303 which came in three types, these were ball which was the common bullet, armour-piercing, and tracer. Rifle ammunition was issued in chargers, each charger held 5 rounds. The chargers with rounds were carried in cotton bandoleers, each bandoleer held fifty rounds, broken down into 10 chargers. Most other S.A.A. ammunition was issued in small cartons or paper bundles to be hand loaded into magazines for Stens, Tommy guns, automatic pistols, or placed into cylinders as with the revolvers.

Machine Carbines and Sub Machine Guns

STEN Machine Carbine 9mm

Designed by Major Shepherd and Mr. Turpin as an inexpensive and easily manufactured submachine gun, this was the British and Commonwealth general issue 9mm SMG This was done so that captured German ammunition could be used if supply ran short. England had to think thriftily in her darker hours. The STEN variants were used throughout the war and below are the model variants. The average weight was 7 pounds, though the MK V and (S) variants will weigh more. Each variant held a 32 round magazine. Rounds per minute fired ranged from 450 to 600 depending on the variant. The effective range was between 50 and 75 yards.

An unofficial sniping team plots out their target. Of interest again is the use of one rank chevron.

The No.4 MK I (T) with two loaded chargers.

STEN MK II

Introduced and first used for the infamous raid on Dieppe August 19, 1942. This variant was the most commonly issued of all that were manufactured. The Sten MK II saw use until the end of the war. It featured an all steel rotating mag housing and perforated barrel nut. The MK II was prone to jamming which I am sure cost many a soldier his life. Should he be crawling and get dirt into it, as the port was fixed open, it would foul. It was very lightweight and easy to breakdown.

Here is a Polish officer with a Sten MK II with pistol grip. Note the private purchase leather gloves, and special Airborne officer's folding map case. This gentleman is wearing a fibre rim helmet – just visible is the rim peaking out from underneath the netting. Also of interest is the pistol ammo pouch on his right, and the binocular case and compass pouch on his left. Note the whistle hanging just underneath the binoculars. He also wears a tie and collared shirt.

Here a Polish trooper prepares to fire. Note how he holds the barrel nut, which is how the weapon was to be held and fired, as if one held onto the magazine it would cause the weapon to jam – this was a well known flaw on this model.

STEN MK II (S)

(S) Meaning Silencer. This variant featured a 'screw on' silencer. It could only be fired on semi-automatic while the silencer was employed.

STEN MK III

The MK III was widely issued to Glider Pilots. It was not a weapon of the Paratroops or Glider Infantry. This variant featured a solid barrel with metal tube body which was welded along the top. There was no foregrip and one had to be very careful to not burn their hand on the barrel. For example the author once, and once is all it took, held the barrel while wearing a glove. I thought this should suffice ... I pulled the trigger and my glove immediately disintegrated. The end result being a charred hand and a very burnt glove.

The author armed with a MK III Sten, the shoulder stock is hidden by the web shoulder sling and high grass. Note the characteristic barrel of this model Sten. Issued primarily to Glider Pilots. Note the R.A.S.C. plastic cap badge on the beret.

STEN MK V

This variant was made in a more conventional manner and featured a wooden butt and pistol grip. It first saw use during the Normandy Invasion of June 6, 1944 and remained in service until the end of the war. Of all the Stens this variant was the most popular as it was the best made and most reliable. This variant replaced the MK II for the Airborne as their general issue SMG. Nine out of ten Stens used from Normandy on were MK Vs. The only real drawback was the added weight of the wooden stock.

Above: A young section leader cries out for help as an enemy tank rumbles towards them. The situation is desperate as behind the tank are infantry and the section leader has but 2 1/2 magazines left. He is armed with a MK V.
Right: A MKV broken down with its ready magazine secured and underneath the parachute harness.

On the left the corporal is armed with a Sten MK II, on the right the author has a No.2 revolver and two enemy grenades.

Chapter Twelve: Airborne Weapons

STEN MK V (S)

Also known as the STEN MK VI. This was the silenced variant of the MK V which had a similar 'screw on' silencer as the MK II (S)

Here is the MK VI or MK V (S) in use by a scouting party. Note the barrel jacket cover which protects the hand from burning. Also of interest are the issue wool cuffs on his smock.

This model of Sten was more commonly used by Commandos. Also seen here is the leg strap of the tanker's holster on his right thigh.

On the right is a S. Staffs platoon leader armed with a MK II Sten with open frame shoulder stock. Both men wear the 2nd pattern Sten bandoleer. Of interest is the beautiful 2nd pattern smock as worn by the Sergeant. Note the button tabs and non wool cuff.

Thompson Sub-Machine Gun .45

Known as the Tommy gun and was part of the Lend-Lease Program. Each Thompson cost £45 Sterling, making it a very expensive weapon for a country whose pocket book was in dire straits, especially after Dunkirk. At the outbreak of the war this was the only SMG available in quantity from the Allies. After Dunkirk there was a need for a light automatic weapon. As the Thompson was put into service the British introduced their own SMG's which were the Lanchester and finally the Sten. The Thompson was used in limited number by the Airborne and then replaced by the Sten as it was much cheaper and available in greater quantities. By the North African Landings the Sten was the common issue SMG. Note that a Thompson in the use of the Airborne from this time and on was very limited and quite the exception. Of interest is that a Thompson cost $225.00 each compared to the Sten which cost £3-5 pounds each. By today's rate of exchange this was about $5.50 to $8.50 each. The Thompson weighed 10.7 pounds and held either a 20 or 30 round magazine or a 50 round drum. The drum was not used by the Airborne. Its rate of fire was 675 rpm. It's effective range was 50 to 75 yards.

Pachett Machine Carbine 9mm

The Pachett gun was developed by G.W. Pachett and was first used on an "experimental/trial" basis by members of the South Staffordshire Regiment, Air Landing Brigade, 1st Division for the Arnhem Operation. Some 100 guns were issued. As there were not too many of the South Staffords who came back to report upon the trials and tribulations of the weapon the experiment was postponed. The gun did not see use again until after the war. The Pachett eventually became the Sterling and was reintroduced in 1953. Each gun weighed 6 pounds and held the same magazine as the Sten which was 32 rounds. Its rate of fire was 600 rpm.

An acting section leader S. Staffs barks out as command to a rifleman. Parachutes litter the ground from the 21st IPC who came in first to mark the LZ.

A close up of the Pachett, a very compact weapon. Note the stamping of the Sten sling above his right hand.

The stock folding into place to then locks into its operating position.

The folding stock of the Pachett. The shoulder rest folds out, then the stock folds out from that. Thus a double folding stock.

Left: Here the section leader is about to lock the stock into its final operating position. Note how the sling is simply hooked to the perforated barrel.

Chapter Twelve: Airborne Weapons

Machine Guns

Bren Light Machine Gun .303

The Bren was the workhorse of the British Army throughout the war and into the post-war years. Being a light machine gun it served as a section (squad) support weapon. In a line infantry section there would be a three man Bren team consisting of gunner, assistant gunner, and a lance corporal, who directed its fire. In an airborne Bren team you would find only two men – the gunner and the assistant gunner. Its weight was 19 to 24 pounds depending on the variant of which there were four, though only the Mk I and MK II saw service with the Airborne. The MK III and Mk IV were introduced too late for wide use in the war. The Bren was magazine fed with each magazine holding 30 rounds and its rate of fire was 500 rpm. However, the time taken to change to fresh magazines reduces the rate of fire to more like 120 rpm. Effective range was 1,000 yards. It was said that, "bullets hold ground not bodies," as the object of a defending battalion is to have so much lead flying across their front as possible. In the reverse the LMG is to provide covering fire against the enemy who may attempt to hold up an attack.

This Bren Gunner stops to catch his breath and takes note of the pillaring smoke from just over the horizon. Note the VI flash on his beret, as he is a member of No.6 Commando. He wears the denim overalls over his battledress. The Bren can be fired in this position or with the bipod folded underneath by gripping on to the bipod for support. Of interest is the canvas Assault Jerkin which is simply a pair of deep breast pockets that could hold Bren or Thompson magazines, and 2" mortar bombs.

A series of shots which show how a Bren barrel was changed. As it could warp from overuse, the barrel had to be changed after so many rounds had been fired. A spare barrel kit was, by regulation, to be carried along with each gun. Shown here, two Paras change their barrel as more rounds than preferred have been shot through the one barrel. Against the fence is the Bren magazine chest which carried 10 magazines. No.1, the gunner, removes the barrel as No.2 is ready with the spare.

The barrel is out and No.2 is there with the spare. Note that there is no magazine in the gun. To remove the barrel it is twisted and turned and then pulled out. No.2 then slides the spare in and will then turn and lock it into position. Note the sliding action cover which is pushed forward, leaving the breach open to accept another magazine. When the gun is in transport this cover is slid back over to protect the 'insides'.

The barrel is in and a fresh magazine must be added quickly.

For King and Country: British Airborne Uniforms, Insignia & Equipment in World War II

Vickers Machine Gun .303

The Vickers was the general service machine gun of the British Army for the Second World War. It was belt fed and water cooled with a rate of fire of between 450 and 600 rpm. Effective range was up to 3,700 yards. It weighed 40 pounds – this weight includes the gun's jacket being filled with water. The tripod weighed 50 pounds.

Anti Tank Guns

PIAT (Projectile Infantry Anti Tank)

"A load of rubbish," remarked Tish Rayner 2nd Battalion, D Coy Oxf and Bucks in reply when questioned as to his thoughts of the PIAT. As a side note Tish was in the sixth glider which went missing June 6, 1944. He and the rest of his platoon arrived at the bridge (Pegasus Bridge) 24 hours later.

The PIAT was a spigot fired anti tank projector. It fired single shot hollow charged high explosive 3.5" bombs which weighed 2.5 pounds each. It took 200 pounds of pull to cock the weapon. Once the weapon was fired it would then self cock. Once it was cocked the bomb was slid into the tray and then fired. This is what saved the day for the lightly armed Paras as the PIAT was all too often their only anti tank weapon available. The PIAT could disable an enemy tank, and blow holes through buildings, which was a preferred method of house to house fighting known as 'Mouse Holing'. It could also fire 2" mortar bombs. A specially manufactured tray adapter enabled the PIAT to launch a 2" bomb indirectly like a mortar making the PIAT a dual purpose weapon. Designed to be used by a two man team, often in combat situations it was a lone Para who braved the odds and dared to 'duke it out' with the oncoming tank. The PIAT weighed 35 pounds and its range was up to 350 yards, yet any veteran will tell you that, "You had to get up bloody close to make it count!" With this in mind the maximum range for anti tank use was 100 yards. The 350 range was more for 'Mouse Holing' or 'House Breaking' as it was called. This was how a house was cleared and or enemy snipers in building were dealt with.

The PIAT bombs were carried in a special cardboard tube container which was hand carried using its web carry strap. Each tube container carried three bombs at one bomb to a row.

A Vickers in use at Driel, Holland. These are Poles dug in and defending the perimeter at Driel. Fighting was desperate as the Poles had few, if any, PIATs to keep the German armour at bay. Note the Gunner wears a 1st pattern Denison, as not only is it very 'washed out', but also there are no brass snaps in the rear of the smock. Of interest are the blocks of earth that have been cut out and formed into a make shift defensive position. As the ferry was gone and no boats were available during the evacuation at Arnhem, the Poles fell back to Driel and the order had only just been given to prepare defensive positions when they were attacked by motorized infantry.

Above and below: A PIAT team. Visible here is the bomb carrier made of thick card and held in place with two wooden braces and metal and web strapping. Note the large earlier war net worn at left, and the smaller more common later war 1944 net at right. Note the bomb tips, this is what triggers the detonator. The tips are simple plastic and once crushed in by the force of heavy impact they explode.

Gunner King Demonstrates the cocking of the PIAT. It took 200 pounds of pull to cock this weapon, and it was not an easy task! Note how his foot is braced on the shoulder stock for leverage, and the jeep trailer in 'Mickey Mouse' camouflage scheme.

Chapter Twelve: Airborne Weapons

6 Pounder

Standard anti tank gun of the airborne forces. Capable of disabling a heavier tank and destroying a light to medium tank. Special modifications were made to the gun for Airborne use, such as the removal of the lower shield, in order to lighten the load while in flight (in the glider).

17 Pounder

Heavier and only able to arrive via the larger glider, the Hamilcar. Able to take on the heaviest tank the enemy could field. 17 Pounders were towed by (transported) 15 CWT Morris Vans (trucks). Both of these were carried in the Hamilcar.

Bombthrowers and Flamethrowers

2" Mortar, MK VIII

The MK VIII Mortar was a specially shortened mortar for airborne use and was 21.37" in length. It consisted of a steel barrel, breach piece, firing lever and lanyard, and spade. The 2" mortar was also known as a 'bombthrower' as it would throw a bomb through the air and onto its target. This was the platoon's light support weapon and its main purpose was to make and throw up a smoke screen that could hide the movement of advancing or withdrawing troops. It was also capable of throwing high explosive bombs to destroy targets that could not be taken out by small arms fire. The number of bombs that could be carried and that might be available was limited so they had to be used sparingly. Parachute flares were also available for use at night. The effective distance of a HE (high explosive) bomb was about 8 yards in open ground from where the bomb landed. Fragments from the bomb may, with enough velocity, inflict wounds up to 150 yards especially if the bomb burst upon rocky terrain. The range of the 2" mortar was from 100 to 500 yards.

Each bomb was similar in construction as each had a body, internally threaded spigot and tail unit. Each is also marked with the contractor's initials or trademark, date of manufacture, and the mark of bomb that it is. There were five types of 2" bombs.

Here is a 6 Pounder in action. Note how the Gunner has his water bottle attached to his web braces, and that the shield has been removed, which was done to fit the gun into the glider.

The MK VIII Airborne mortar, from the left is the spade, cover, behind the cover is the breech piece which houses a spring loaded plunger that pivoted on a stud in a recess in the breech piece. The plunger bearing on the upper side of the recess ensures that the catch functions properly. The underside of the catch is serrated in order to engage the serration's on the barrel. To the right of the barrel catch is the flat steel firing lever with lanyard which is made of two 12" lengths of 1/4" plaited line with its ends secured by whipping. The barrel and sighting line are in white which acted as a crude site for aiming the bombs.

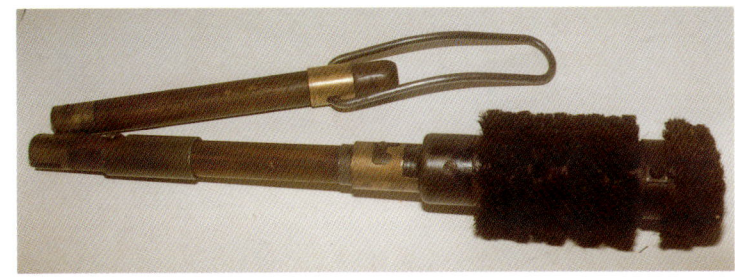

The barrel cleaning brush. As the barrel was fouled from use it had to be scrubbed and cleaned out. There is a 2" mortar P-37 web cleaning kit carrier – this brush was carried in it.

2" Bomb Carrier

Two inch bombs were carried in a special cardboard tube container which was hand carried by its web carry strap. Each tube container carried six bombs at two bombs to a row. Each carrier was marked with a colored stripe to identify its character. The colored identification stripes were: yellow for HE, green for smoke, and white for parachute flare.

High Explosive Bomb

HE MK I consisted of a body made of solid drawn steel with parallel walls that were rounded at the base and formed a central spigot which was screw threaded to receive the tail unit. The tail unit had six vanes (fins) and a central tube and cartridge retaining cap. The central tube was screwed into the central spigot. A grub screw then secures the tube. The bottom is screw threaded internally to receive the cartridge retaining cap. Eighteen holes of six rows of three are spaced between the

Left: From the bottom up is: the spade, the breech and to the right is the cover, lanyard, firing lever, the barrel catch is hidden in this shot – just beneath the sighting line (the sighting line is simply painted on the barrel.) To fire, one pulled the lanyard once the bomb was dropped into the tube.

vanes to evenly distribute the gas pressure around the bomb. The retaining cap has a central firing hole to allow the firing pin to fire the cap of the cartridge. The bomb is filled with 5 oz. 13 drs. of Baratol 20/80 which is surmounted by a millboard washer and a wax paper collar to receive the base of the fuzz. The HE bomb weighed 2.4 pounds and had a red band painted around the it, indicating that it contained an explosive. It was then stenciled in black over a green band BAR 20/80 or AMOTOL 80/20. Overall the bomb was varnished yellow for further identification.

Smoke Bursting Bomb
The Smoke Bursting bomb MK III consisted of a body and steel tail unit. The Smoke Bursting bomb was filled with Phosphorus and varnished green with a red band to indicate an explosive and then a white band to indicate smoke. It weighed 2.4 pounds.

Smoke Bomb
The Smoke bomb consisted of a body, tinned plate cup, adapter and closing disc and tail unit. It was filled with Composition SR 269 for the MK II or PN 303 Composition SR 269 for the Mk I bomb. The body was made of thinned steel and closed at the top by a nose cap which was sweated and soldered on. It was varnished green and had a red band to indicate an explosive. Stenciled below in black was 2" MOR II, beneath this in black was S.R. 269, then beneath that in black P.N. 303. This bomb weighed 2 pounds.

Illuminating with Parachute Bomb
The Illuminating with Parachute bomb MK I consisted of a body, delay holder, adapter and tail unit. The body was made of tinned steel and closed at the nose with a tinned plate lid. It was filled with a star unit (flare) and parachute. This bomb was left unpainted and simply featured it's particulars stenciled in black ILLG. WITH PARACHUTE. It weighed 1.1 pounds.

Signal Multi-Star Bomb
The Signal multi-star white MK I bomb was similar in design to the illuminating with parachute bomb in the assembly of the delay holder, adapter, closing discs, tail unit and cartridge. A colored band above the stenciled lettering MULTI indicated the color of star contained. A red band above the word MULTI with a green band below it signified that the bomb was filled with red and green stars. Each bomb was filled with the particular colored star composition and weighed 1.15 pounds.

Signal Single Star Bomb
The single star bomb was identical to the multi-star with the exception that it contained only one star.

3" Mortar
Introduced in 1938 and used throughout the war, the mortar, besides the 75mm Howitzer, was the heavy support weapon of the Airborne. The 3" mortar fired a heavier bomb to a greater range which therefore caused more destructive damage than the 2" mortar bomb. The 3" mortar had no trigger or striker like that found on the 2", and it featured a fixed striker pin at the base of the tube. The 3" bomb had a percussion cap on in the base of the round and is fired when the bomb is dropped down into the tube. The cap hits the striker and this causes an explosion which send the bomb on its way to the target. The 3" mortar weighed 120 pounds and when possible was carried in a 15-cwt truck, jeep or Bren carrier. The weapon was broken down into three pieces to be man portable, and consisted of the tube, base plate, and bipod.

The 2" bomb carrier (this carrier was found in Arnhem). Note how the type of bomb is identified on the carrier, and the web band carry strap is visible at the middle bottom. Each tube is separated by a block of wood and secured by metal banding, and each tube cap is attached to the web tapes (strapping) which go completely around the carrier. As can be seen it's simple heavy card stock and it is surprising that it has lasted this long.

An HE bomb void of any markings or paint as it was dug up in Arnhem as these have not survived the elements.

Here the safety cap has been removed and the fuse is exposed. Just visible on the top of the safety cap it reads, REMOVE BEFORE FIRING.

The SMOKE bomb. Note the stenciling is nothing like that found in the manual yet it is varnished green, with a red band, and in black you can see S.R.269.

The ILLUMINATING WITH PARACHUTE bomb.

3" Bombs
The 3" bomb has a streamlined body with a screw on nose piece (cap) which is the percussion fuze. A perforated container filled with the propelling charge is secured to the base. Vanes are attached to the base to keep the bomb steady in flight. A bomb was either filled with HE, smoke, or star composition. Each bomb weighed approximately 10 pounds each.

Chapter Twelve: Airborne Weapons

Above left: Here members of the Polish Brigade are on exercise with the 3" mortar. All wear the 1st pattern jump helmet, the fiber rim, as well as their modified Irvin Flight Crew jackets. The tell tale sign that they are Poles are the collar kite badges, and the shoulder title of the man on the far right, "POL", the "AND" being hidden underneath the jacket.

Above right: The Poles at a demonstration with dignitaries and officials present. The P-37 web tubes are for carrying 3" mortar bombs. This was a sort of bomb jerkin. Four bombs were carried on the back, and three on the front. This must be a mortar section, as the man in front of the line is holding the base plate, second in line the bipod, and third in line the barrel, and the remaining five men carry the bombs. Of interest is the fashion in which the men are wearing their haversacks. Also note that all E tool handles are the earlier variant.

Right: Here members of a Support company are moving their 3" mortar to a small patch of ground across the street from a residential area in Arnhem. The trooper closest carries the barrel, behind him the ammunition bearer, and lastly the base plate.

3" Bomb Carrier

Three inch bombs were carried in a special cardboard tube container which was hand carried by its web carry strap. Each tube container carried three bombs at one bomb to a row.

Flamethrower

Manufactured by the Petroleum Warfare Department. Known as the "Lifebouy" as it resembled a child's rubber bathing ring (beach toy). It had a range of 50 to 60 yards and weighed 55 pounds when full of propellant which was a mixture of gasoline and nitrogen. The Marsden portable Flame-thrower was first introduced on July 3, 1940, and could fire a stream for 15 seconds continuously, after which it would have to be refilled or would be abandoned. The fuel capacity was about 25 gallons. Not much is known about this weapon or where it was used. I believe that Sappers used them at Arnhem in limited number.

Artillery and Anti Aircraft

75MM Howitzer

Lend-Lease from the U.S. This was the light artillery of the airborne forces and widely used throughout the war, and was one of the few saving graces at Arnhem. The

Right: This is an interesting photo in that the gunner leaning over the 75mm is wearing a No.1 bayonet for the No.1 MK III Rifle. Prior to the Poles becoming part of the 1st British Airborne Division, they were not well equipped and especially not with the latest weaponry. In the middle behind the gun crewmen are General Sosabowski and General Sikorski. Of interest, the two officers first and second on the right are wearing their metal rank stars on their Denison shoulder straps, this is most unusual, as rank was for the norm not worn by the Poles on their smocks. The main focus of this shot is truly the 75mm which was the workhorse artillery piece of World War II. The British did not paradrop them like the Americans, but preferred to bring them in by glider. Just visible are the Polish eagle stencils underneath the helmet nets on the trooper on the far left and the gunner in front of General Sikorski.

British did not dismantle and drop these by parachute as the Americans did, they preferred to keep them assembled enabling the gun to go straight into action by glider.

Polston Anti Aircraft Gun

The Polston's design originated in Poland, and after the country was overrun in 1939 the Poles sent it to England. The Polston was considerably cheaper than the already in use Oerlikon anti-aircraft gun. The first Polstons were delivered in March of 1944. Two of these guns were used at Arnhem by the Recce Squadron. Arnhem seems to have been the proving ground for unusual and trial weaponry. The Polston fired a 20 mm round from a double stacked 30 round magazine. It weighed 121 pounds and was 86" in length. The barrel was 57" long and had a rate of fire of 450 rpm. Its maximum effective range was 6,630 feet.

A 75mm in tow. This is truly an early shot as the soldiers are all wearing the fibre rim helmets, and are armed with the No.1 MK III rifle. The PL is the Polish vehicle identification marking found on all Polish transport.

The Polston gun, though in this shot in use by a non airborne soldier. His boots do not look British, yet are possibly American or even the special Canadian Invasion boots. Note the spent magazine (drum) cast to the side to the left of the gun. As mentioned the Recce Squadron had two of these in Arnhem.

This is the No.36 Grenade. On the right is the break away for detailed listings: from the top is the striker, to the top left of the right bomb is a screw hole. This is called the filling hole and is closed and secured by the screw plug which is not shown. The dark green areas between the red and the center is where the explosive filling is stored. To the right of the center is the detonator which looks like a brass tube. The large screw like cap at the bottom is the base plug. On the outside is the safety pin, and just underneath that is the striker lever, known by the Yanks as the 'spoon'.

Grenades and Mines

No.36 Bomb (Mills Bomb)

General issue and most commonly used grenade for the British and Commonwealth Armies. Useful in clearing out dugouts and buildings. Can be thrown from 25 to 35 yards away and weighs 1 1/2 pounds. Perfect for street and close quarter fighting. It consisted of a cast iron body which was filled with high explosive. In the middle of the body was a center piece which held the striker and spring. These are held in place by a lever inserted in a slot at the top of the striker. The lever is secured by a safety pin. When the safety pin is released the striker acts on a percussion cap and a length of safety fuze. After an interval of time the grenade is detonated.

No.69 Grenade

The 69 was made of bakelite plastic and was filled with high explosive. It served as a light weight percussion grenade able to demoralize the enemy by the shock of the blast, or to throw it behind enemy troops hiding in bushes or behind a bank rather than in front of them or at advancing troops. It weighed 3/4 of a pound.

Right: On the left is the No.36 grenade, and on the right is the No.69 grenade. Note the screw plug on the No.69 which is much like the one found on the No.36. This screw is removed and the filling poured in. The red XXX is an identification for TNT. Visible at the bottom of the No.69 is its base plug.

Chapter Twelve: Airborne Weapons

No.73 Anti-tank Grenade

Designed to disable an armoured vehicle by blowing off its track. Its range was between 10 to 15 yards as it weighed 4 pounds. Because of the high explosive encased within the bomb and its limited range, the thrower had to be behind cover. Consisting of a tinned plate casting filled with high explosive, it had the same mechanism as the No.69 which was covered by a safety cap.

No.75 and No.75A Grenade

Of U.S. manufacture and designed to disable a tank by blowing the track off or damaging its suspension. It was detonated only by the tank running over it and crushing it. It could be thrown in front of a tank or laid out and covered in the dirt – it could also be used to destroy trains. It weighed 2 1/4 pounds. The No.75A has an 'A' stamped on to it and has only 80% the power/blast of the No.75. The thin tin body was filled with high explosive and on one side are two pockets with slots cut into them. These slots hold the detonators. These pockets can be closed by folding the metal sides over them. The striker plate is mounted above the detonators by two brackets. The cap at the end is where the HE was filled into the bomb and must be kept unscrewed. The cap is cemented in place.

To the left is the red XXX TNT identification. At right is a cut away for a detail listing, from the top: the lead weight which is attached to the white tape, underneath the tape is the striker and creep spring, beneath this is the cap pellet and cap. The brass tube is the detonator, the yellow area represents the HE filling, to the bottom left is the screw plug where the HE is filled through, and on the bottom is the base plug.

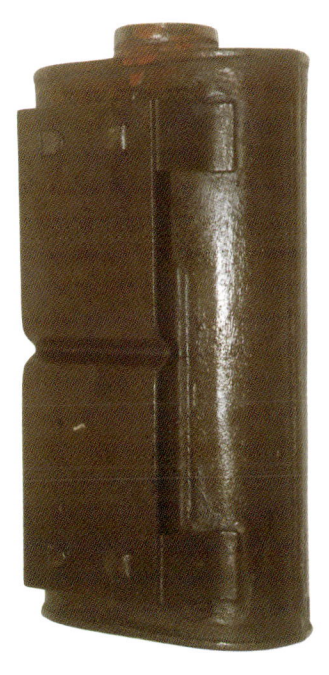

Left: Stamping detail of the No.75 grenade. Right: The striker plate which the tank will crush and detonate the bomb.

Here a Lance Corporal counts his blessings and holds his breath as an enemy tank rumbles towards him. At the right moment he'll place the bomb where the tank will surely run over and crush it, thus setting off the explosive. The idea of the No.75 grenade was to stop a tank by breaking their tracks or causing some damage to its suspension system. The LCpl. is armed with the MK V Sten, and wears a fibre rim helmet.

165

No.77 Smoke Grenade

This was for quickly producing a local smoke screen to mask an attack or withdrawal. The grenade is exploded by a detonator which produces smoke instantaneous. It weighed 3/4 pounds, and consisted of a tinned plate casting with a screw on tin plate lid. The detonator is located in the center of the top of the body. The body itself is filled with phosphorus.

The S.T. Grenade

The 'Sticky Bomb' was designed for use against AFV's (Armoured Fighting Vehicles). It would stick to the target thus exploding and damaging the AFV. It would not however 'stick' to sloped, wet, muddy or oily surfaces. Perfect for dropping onto an AFV from an upstairs window like a little bundle of joy for you friends of international flavor. The body consisted of a flask filled with high explosive. Prior to use the grenade was protected by a special hinged flask which the grenade was carried in. The inside of the flask was covered in a sticky substance to keep the bomb from moving inside. The detonator was contained inside of the handle of the grenade. Also inside of the handle is the striker and striker spring. A small safety pin was also found on the handle which secures the spring and lever. Of interest a label is attached to the safety pin and it reads "Warning do not remove this pin until ready to throw grenade." The S.T. Grenade weighed 2 1/4 pounds.

The No.77 Smoke Grenade. From the top: safety cap, lid, adhesive tape will be found stuck to the cap and the body as a precaution, and lastly the body.

From the left: throwing handle which houses the striker and spring, towards to the top of the handle is the safety pin, then the rubber screw ring which holds the bomb together and secure. The ball shaped object is the sticky envelope which contains the glass flask and HE filling. Surrounding this is the metal protective casing which is removed when the bomb is to be used.

No.82 Gammon Bomb

The charge was variable according to the job required as one added the explosive into the cloth bomb container. To arm and use, the bakelite cap was unscrewed and a white tape with a lead weight on the end was thrown at the target as the lead weight uncoiled the tape and pulled out the safety pin thus arming the fuse. Once this fuse was ignited the bomb would explode upon impact. A special plastic explosive capable of destroying an enemy tank, by German records, this little gem destroyed 85% of its armor at the battle of Arnhem.

Below: The L.Cpl. is unscrewing the safety cap of a No.82 Gammon grenade. Once the cap is off he'll throw it at the enemy vehicle. Note that the wool cuffs on the Cpl.'s smock are not issue and have been added later.

The detonators box which is kept separate from the No.82 grenade, as found in Arnhem.

Right: The No.82 grenade. From the top: safety cap, adhesive tape for added safety, filling hole, and envelope where the plastic is held.

Chapter XIII

'Attached' Troops of the Airborne

Each Airborne Division had specialists attached for specific duties such as liaison officers, working between the Paratroops and the seaborne Commando who are to link up at a predetermined location – Pegasus Bridge for example. Members of X troop and No.10 IA (inter allied) were employed as scouts, as they knew the lay of the land and as interpreters and were able to communicate with the local populace and interrogate prisoners. This came in very handy for Operation Market Garden. In this volume I will only delve into the British, Dutch and X Members. Future volumes will cover the remaining nationalities whose countrymen earned their wings and served with distinction.

This shot was taken November 1, 1944, just after the raid in Flushing, Walchern, Holland. Here are members of C Troop, No.4 Commando with a piece of German booty. The officer on the far left wears the V neck sweater, and the rest wear Denison smocks. By this stage in the war the Denison was of general issue to the Commando. Of interest are the regimental cap badges of the trooper's former regiments such as the Green Howards, K.O.S.B., Cameronians, and the R.A. Also note the knit cap comforter worn by the man on the far right, last row back. Shell dressings are seen tucked into several of the Denison shoulder straps. This photo was donated by D.A. Lambley of No.4 Commando.

Those Who Served
George Jones C Troop No 4 Commando

My name is George, I am 75 years young, I walk at least 2 miles a day with my dog Foxy, a Fox Terrier. I worked in the building trade before joining the army in 1939. I was posted to the Royal Tank Corps (RTR, Royal Tank Regiment). Did six months with them, then volunteered for the Special Service Brigade (Special Service Battalion) now the SAS. After parachute training, left and helped form No.4 Commando in Weymouth 1940. I was on the first raid on the Loften Islands in Norway. Come back to do many small raids on the French Coast. Left for Africa for special service work Training Africans to pilot landing craft for a special landing. This was however not put into service. Returned to take part n the Dieppe raid with the Canadians. Landed D-Day before the main force on Sword Beach to put 6" shore guns out of action to help cut down casualties of the main force on landing. Moved along the coast to help the French section of No.4 whose job it was to take the Casino, where fighting was very heavy. Lost many men. Day two moved inland to help 6th Airborne to hold Pegasus bridge. Had to street fight all the way, Germans counter attacked many times but we held firm till main army relieved us. Lt. Thomas who jumped and was assigned to 6th Airborne (June 5, 1944) as Liaison Officer for No.4 was killed by his own men as he forgot the password. His batman Sam Ryder ended up with the Canadian Paras as he got cut off from the Commandos during the fighting.

In C Troop we were #1 in rock climbing, parachuting, marching and fitness. But the rest of the Commando were very good. We had four Americans with us on the Dieppe raid. One a Cpl. Bowles, won the Military Medal. They belonged to the Texas Rangers (The U.S. Rangers were formed after and inspired by the British Commando. Their first bloodletting was at Dieppe).

R.J. Anderson C Troop, No 4 Commando

I joined up in 1939 and was posted to the Royal Ulster Rifles in Ireland where I spent a year. Then a notice came 'round asking for volunteers for a Special Service Regiment I applied and was accepted and went to Scotland where I did my Commando training. After this our name was changed to 'Commandos' and we were No.4 Commando. This was split into troops of 45 men each and mine was C Troop. We lived in 'Civie diggs' (Commandos were permitted to live off base with Civilians as borders, and given a special allowance pay for their keep) – Good ol' days!

We volunteered for anything that came along such as parachuting at Prestwich doing 8 jumps each, 4 from a balloon and 4 from a plane. There we practiced stiff rock climbing in Wales at Betlesda. I was in the Normandy landings and it was pretty rough. Still it was no worse than the others got. We had a few battles near Caen. We also made the raid on Westkapelle, Walchern. I managed to survive and I am now 75 years old and of course cannot remember everything except of course the main things that happened.

Here are C Troop of No.4 Cdo. In Barr, Scotland. Second from the left on the top row is Cpl. George Jones. All men are in shirt sleeve order and wear the wool knit cap comforter, and each man is armed with the No.1 rifle. Note that the man to Jones's right has a collar on his shirt. Is he an officer? This may be a private purchase shirt as Jones did say that they did wear ties in 1942. One last note is that the gentleman in the middle is the only one to wear the green beret with his parent regiment cap badge, and is wearing a collared shirt with shoulder straps (the trooper next to George Jones has no straps on his shirt). (Photo courtesy of George Jones)

A group of C Troop No.4 Cdo., and on the bottom front row second from the left is Mr. R.J. Anderson. All wear the battledress serge blouse and trousers. This photo was taken prior to Dieppe. (Photo courtesy of R.J. Anderson)

Chapter Thirteen: 'Attached' Troops of the Airborne

Jumped Trained British Commandos

One Troop from each Commando of the 1st Special Service Brigade were Jump Trained. This included: C Troop No.4 Cdo., 6 Troop No.6 Cdo., 3 (P) Troop No.3 Cdo. Also included were E Troop 45 RM Cdo. (Royal Marine Commando). Commandos jumped into Normandy as liaison troops with 6th Airborne Division. One such liaison officer was Lt. A. Thomas of C Troop No.4 Cdo., who was killed by friendly fire as he either forgot or neglected to give the password.

X Troop and No.10 Inter Allied Commando

Members of X Troop were made up of escaped German and Volksdeutsch Jews, who were of great service as they served as interpreters and performed "specialist" duties. If caught they would of course be shot as spies. Each man after induction into the Commando, was given an English name and persona to help reduce the risk of execution if captured. Though after Hitler's Order #42 stating that all British Commandos were to be shot on site, it was a 50/50 chance if you survived to be sent to a POW cage or into internment.

Fourteen Dutch Commandos of No.10 IA went to the Netherlands for Operation Market Garden. They served as interpreters and guides to the British and Americans. Twelve were attached to the British. Ten of these twelve actually landed at Arnhem by glider. Two Dutch Commandos jumped with the Americans and funny enough, though not to them, none had any parachute training prior to the battle.

Clothing and Equipment

Clothing and equipment for the airborne 'attached' commandos were the same as their counterparts in the British Airborne (and U.S. Airborne). The exceptions being: the green commando beret, unit insignia on their Battledress, and jump wings, which were not commonly worn on the Denisons of the Commando. An airborne helmet was worn for the jump and once on the ground the beret was put on.

Insignia of the Commando

Cap Badges

British Commandos wore the green beret devoid of badges. An exception being those who 'still' donned their 'Parent' regiment's badges. By this stage of the war this practice had been outlived but in the early years it was quite common.

Right: A photo of a group from a group of Commandos taken at Achnacarry, Scotland prior to the Dieppe raid. From the bottom left: Maxie Cook, Sid, and Fred Gooch. Standing on the right is Johnny Ainsworth who later became a glider pilot and was awarded the Military Medal. The tent is known as a Bell Tent.

Above: An officer scouts ahead as his troop awaits the signal for attack on German positions at Lake Commachio, Italy, April 1945. Note how the camouflaged face veil enables the wearer to blend into the surroundings. This is a typical Commando look, featuring Denison smock, denim overall trousers, and light order of webbing.

Left: A group of Dutch Commandos of No.10 IA Commando. Note the Lion of Orange cap badges on their green berets. Most Co Ops badges are cut round, while the gentleman at front center with the No.4 Rifle wears the uncut Co Ops Tombstone as issued on his early battledress serge blouse. Shoulder titles are worn higher than usual, as there is commonly about 1/2" between the shoulder strap and top of the title. Notice that the man raising his water bottle is wearing a KD shirt, as is another towards the rear, on the left bent over with his No.4 rifle. Note the various berets and their sizes – as can be seen, the war time beret was much bigger in the crown than those worn today.

Right: Here the Dutch Commandos are seen on parade. Note the printed Co Ops and printed titles worn. The flash is the Princess Irene Brigade flash. The Commando standing has the special FS knife attachments on his trouser leg. Note that the man being decorated does not have the same attachments on his trousers.

The brass Lion of Orange cap badge.

All ranks of the Dutch, No.2 Troop, No.10 IA Cdo., formerly of the Princess Irene Brigade, wore the brass Lion of Orange. Some backed their badges with black felt. There were some higher ranked officers who wore a Princess Irene badge of silver wire (bullion).

X or No.3 Troop wore no special distinctive badges as they were a secret unit of escaped Jews, Hungarians, Czechs, Greeks, German and Austrians. Some wore the plastic economy or brass General List cap badge of British manufacture.

Flashes, Titles and Jump Wings
The British wore the following badges: No.4 and No.6 did away with the Combined Operations badge and simply wore their shoulder titles on each shoulder and parachute wings on the right arm below the title. Some old salts could have worn the earlier Jump Qualification badge (light bulb) but this is speculation. No.3 Commando wore the parachute wings between the shoulder title and the Co Ops badge only on the right arm. No.45 Royal Marine Commando did something very unique. On the right arm they wore the regulation lines of three tapes: Royal Marines underneath this was Commando and underneath that 45. Here's where it differed. The Co Ops badge was deleted on this arm only, then the parachute wings were worn. On the left side it was as per regulation with the Co Ops badge.

The Dutch wore the No.10 Commando title on both sleeves, then the Lion of Orange on the left sleeve, underneath this was worn the Co Ops badge, on both sleeves. A smaller Co Ops badge was made for NCOs, as there would be little room for rank chevrons with all else that they had to wear. Jump wings are not in evidence. X Troop wore, as a rule no Commando titles or Co Ops badges as a security measure.

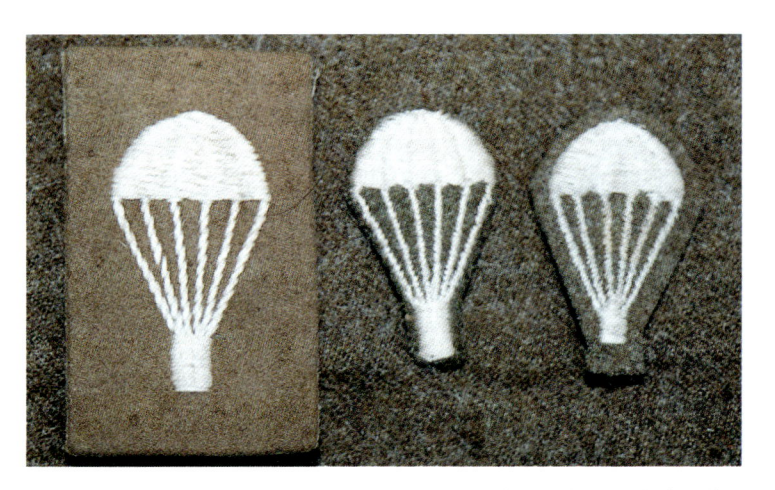

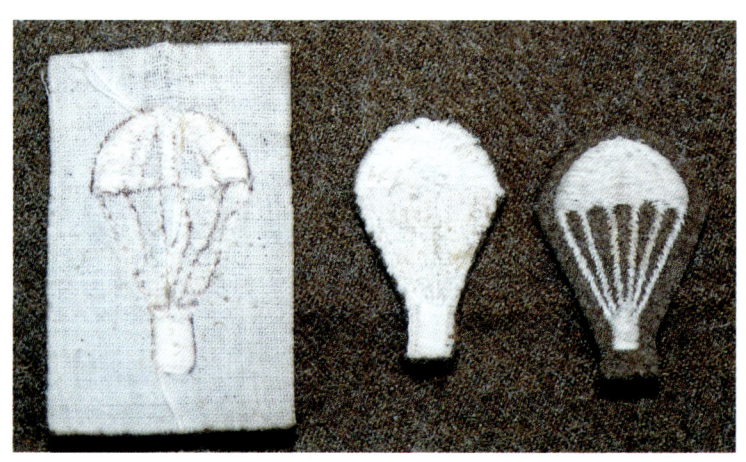

Left: Jump Qualification badge and variants – variants are simply by manufacturer. The shroud lines on the larger badge seem to be 'lined and couched'. The middle and right badges have been cropped which was more commonly seen worn than the square or rectangle as 'uncut' issued badges. This seemed to be the preference of the wearer. Right: Reverse of each badge – all differ by manufacturer's method.

Chapter XIV

'Tobie Ojczyzno' - The 1st Polish Independent Parachute Brigade

Jump training for the 1st Polish Independent Parachute Brigade took place at the main training center at Largo House, located in Scotland. The Poles could well have been the first to use the 'jump tower' as they, like the Americans, (Coney Island Amusement park) had a tower for sport and amusement in Poland, 1937. In 1941 they built one in the UK. This was the first of its kind in the UK. The British were utilizing other methods of training at Leven where their (Poles) HQ was situated. This tower was a 'free fall' tower where the jumper would be carried up to the top and released, descending to the ground below.

These men were trained so intently and for a purpose that, as the years rolled on, only drew further away. For their one and only combat operation, a worse situation could not have presented itself. In the havoc and confusion of Operation Market Garden, such a proud and strong body of men were wasted as if casting small twigs upon the surf. Topping it all off, to be shrugged off with disparaging violations of character cast upon their namesake. Only few know of their true performance and triumphs.

This photo must have been taken just prior to the Poles' attachment to 1st Airborne as each man is equipped with the Mk II Sten. I would think that if they were equipped by the British with a priority, then they would be armed with MK V Stens. The Poles were not well equipped prior to their attachment to the British and were using outdated equipment. Here General Sosabowski and an English general staff officer review a Signals or Reconnaissance platoon. Note that each man who has a Welbike also has a folding airborne map case, and may be part of a Recce or Signals unit. Just after the last Welbike trooper is a signaler with a No.38 set hung from his breast. Also note that each man seems to be wearing the fibre rim jump helmet.

Those Who Served
Captain Jan Lorys 1st Polish Independent Parachute Brigade, G2, Headquarters Company

Leaving France, (Members of the Polish Army escaping with the BEF as France fell to the Germans.) we arrived in the UK. From there we were transported to Scotland (by train at night). There I was assigned to the 1st Brigade, which was commanded by Col. Sosabowski. The 1st Brigade was then renamed the 4th Cadre Brigade. This brigade Sosabowski managed to move to Eliock in Lanarkshire, Scotland. Here we lived in tents. There were about 300 of us and many, too many officers – as we were a cadre brigade (troops which would be used to build up other formations). We had quite a nice location there, though with the coming of winter we were moved to better quarters at Fifeshire (Scotland).

It was so cold at Eliock that when I wanted to have a shave, I would go to the brook and break off the ice to get some shaving water. We remained there until October. While still stationed at Alyook, I was sent to take part in a shooting and mining course. This was the first 'special' course to prepare us to be sent to Poland as Commandos. This course taught you how to use the various weapons (Allied and Enemy), how to use various mines in blowing up railway bridges and machinery in factories and also we were trained in hand to hand combat, including Jujitsu. Two Jujitsu experts from Hong Kong were instructing us (Fairbain and Sikes! I asked Major Lorys if this was indeed the infamous Honk Kong Policemen. He said, "I did not know them by that name. We called them John and Peter."). These were the first steps in training the Polish troops for use in Poland. Col. Sosabowski was in command of this group, who later learned from General Sikorski, that there was some consideration in training some parachutists. Sosabowski was very interested in these parachutists.

Parachuting was very popular in Poland before the war. Considered sport and that of a hobby, parachute jump towers were built, much like those at Coney Island (NYC, USA). Though the Polish had built theirs first. In years to come, the Poles again built one in the United Kingdom. This was the first of such towers in the UK at that time. This particular one was built near Largo House which was the Polish parachute training center. The Monkey Grove at Largo House was famous for its obstacle course.

Our training in 1940 and the beginning of 1941 was somewhat unusual. We were a Cadre brigade with a large percentage of officers. We did not need to start training from the bottom but once we started getting recruits, the order of the training looked like this: Basic military training-discipline, weapons knowledge, marching, some shooting. Largo House preparation for parachute jumps-which were from the parachute tower at RAF Ringway we made jumps from balloons and airplanes. Those who completed these jumps satisfactorily were returning to their units to continue proper parabattle training, and here because we had no planes, we had to use lorries to take troops into new unknown to them fields or villages to do their attacking, marching and or preparing defensive positions. And then we were short of military equipment, ammunition, etc.

Largo House was where we did our parachute training. As we did not have enough planes, we sent men out in wagons (trucks) and they would jump from the back of the wagon and begin simulating hard fighting exercises, attacking, and defending. There was normally another unit that would take part as an adversary. We had night attacks on buildings, and various defensive positions. Saturday we usually cleaned our weapons, and on Sunday we had a free day after services (church service).

We had a special place where we could do some running and some jumping to keep us in a top physical condition. Headquarters delegated how many hours each unit was to spend on weapons training and or what exercises to do. Each unit had a weekly program (as passed down from brigade HQ). Artillery, machine guns, and mortar training were more in theory (how to take position, how to change position and how to stand and shoot) as ammunition was in very short supply (they had to pantomime). We were not issued with flame throwers. The British told us that they were short of these things, so we did not get them as we were not being trained to serve with the British but rather to fight in Poland.

Our daily program went something like this: Get up 6 am (7 am in the winter) have prayer, wash, breakfast (meat was rationed, we received one egg a week), PT (Physical Therapy [no PT on Sundays]), and various military exercises: 15 to 20 mile march (weekly march), half hour to two hour weapons classes and field exercises, and specialty courses (instruction and application of specialized equipment's for example). Every day we were training with weapons, shooting and exercises.

We were supplied with British food (rations), and we were having our own cooks who prepared the food the best they could for us. Our rations were restricted in what we received, yet it was good food. I remember once when we received a lot of Poles from Russia who were needing more food than was allowed for them (as they had been starved by the Russians while in captivity). So the British then doubled their rations to fill up their stomachs. The British were always watching and a few months later they noticed that some bread had been thrown into dustbins (trash cans), as they did not need as much food, so the British were very clever, they did not say anything, they just started cutting their rations by a quarter here and a quarter later, then we were all getting the same food amounts.

On our first public jump, September 23rd 1941, many dignitaries were there and General Sikorski made a speech to the troops, saying that 'from today I change your name to the 1st Parachute Brigade', he did not say Polish Brigade as we were all Poles. Prior to this the brigade was know simply as the 4th Cadre Brigade. So this was the beginning of our Parachute Brigade. He also mentioned that our main aim is to get to Poland, to fight in Poland. That is what we are supposed to be trained and used for, and that we were directly responsible to him (General Sikorski). At this point we had some 500 trained parachutists, and from then onwards we started getting more soldiers and training them for proper fighting. All this time we were told (by General Sikorski) that we would be the first into Poland by the shortest way. This became our motto, 'The Shortest Way!', this was our aim all the time, to get to Poland by the shortest way. And as we trained from 1942-44, we were getting more soldiers and we were doing more and more training, but we were always short of military equipment. We were only allocated few equipment's and courses for jumping at Ringway (RAF Ringway, where most of the British were trained), but we were getting on (getting by).

By 1944 we were over 2,400 in strength, on the 3rd of March of 1944 the British authorities started pressing the Polish Government and Commanding General of the Polish Army to be attached to the British Airborne, for use on the continent of Europe. Eventually both our government and General Sosnkowski

Chapter Fourteen: 'Tobie Ojczyzno' - The 1st Polish Independent Parachute Brigade

(General Sikorski's replacement after he was killed in a plane crash in 1943) had to give in as the idea of liberating Poland was getting further and further from us and the Second Front was getting closer (The Invasion of France).We thought that by fighting in France we would get more combat experience for when we did get a chance to fight in and for Poland. On the very day that the second front opened up, June 6th 1944, our government and General Sosnkowski agreed to transfer the Parachute Brigade to British command. Then from that point onwards, we started getting everything we needed. New equipment's, perfect new radios, everything that we were not able to get while we were in Scotland. As many as we wanted we got from the British-and we used them, all.

Our radios were outdated, yet in 1944 when we were attached to the British we received everything and more ... and more. Our radio operators did not know how to use the new equipment, they had never seen them and had to learn how to use them. We were glad to be getting our new equipment's because we had been wanting them for a long time. After General Sosnkowski had made the agreement with the British authorities to attach the Polish Brigade under British command, we had more equipment than we needed, we had everything in excess and supply.

On the 15th of June 1944 we were gathered together, where we received the regimental colors, which was made by Polish Ladies of Warsaw, back in 1942. It was hidden in various places (in pieces) and was secretly sent to England in 1944. It was then presented to us and on that day Col. Sosabowski was promoted to the rank of Major General, it was a very important day. After this we were transferred to England from Scotland, mainly to the Stanford area. Stanford was our new HQ. I know this area because I was with HQ at that time. I came to the Parachute Brigade in 1943 after I finished the Officers Staff school. At that time I was with G2 at Headquarters. We again started training and already there were talks from General Browning (Commander and Chief of the British Airborne Forces) that we were to be used as soon as possible

On the 3rd of July 1944, General Downing, who was at that time Chief of Staff for General Browning, came to the Brigade to ask General Sosabowski if the Brigade was ready to go into action. Of course we were not ready to go into action as we had not even had jumps as platoons. We had only had jump as 10 or 12 people at Ringway, as we had no aircraft to jump as a company and or battalion. Sosabowski said that the brigade could not do it and Browning was very unhappy about this. Browning wanted, at certain times for Sosabowski to command one of his British Brigades, but Sosabowski turned him down, as he wanted to command his own Polish troops. Sosabowski was not interested in moving up in rank, rather he wanted to command a Polish unit. Browning gave us a months respite, so we were able to train and be ready by the 1st of August.

From July to 1st of August we were having a lot of jumps. We started with company jumps, and then battalion jumps and finally one jump with the entire Brigade. This mass brigade jump was named Cora. Unfortunately on one of these exercises we lost 26 people because two aircraft collided in the air. It so happened that I saw this accident. As we were told to be ready, 'Battle Ready' by the first of August, we were ready. And we were soon getting orders for various operations from General Browning. During August, we got five such orders (for operations) which we prepared for and we were ready, on the airfields, but we were coming back as the operations were canceled (as the Allied advance progressed rapidly). We were planning these operations a week in advance and even this was too late as the Allied advance was coming up so rapidly.

On September 10th we got orders for an what was called 'Operation Market Garden'. We planned and prepared for this one like the others – very carefully. We were then put under the command of General Urquhart, Commanding Officer of the 1st British Airborne Division, and were obliged to carry out what ever orders he gave us. We were allocated 114 Dakota Airplanes and 45 Gliders, and were told we would be jumping and landing on the 19th of September, as the British went in on the 17th of September. Some of our officers were sent in as Pathfinders on the 17th, also some of our gliders landed on the 17th who were liaison units to the British (there was also the Anti Tank sections, armed with 6 Pounders, who landed via glider prior to the main Polish drop-on the North side of the river. The main Polish drop was to be on the Southern side of the river, as planned by the British). On the 19th we were on the airfields and we waited all day, and were later sent home as the weather was bad (these troop transport cannot fly in dense fog and or heavy rain).

On the 20th the same thing happened and we did not go. And on the 20th we received a change of orders, at this time we did not know why, though we found out later that this change was because of the bad situation with the Division. General Urquhart sent us an order to change our DZ, which was originally right across on the other side of the bridge (main bridge, as for those who do not know, there was also a railway bridge and a pontoon bridge). Our new DZ was to be in Driel (across the river but further down and away from the bridge). We were then to jump on the 21st, the weather was still not perfect but still we went to the action. We were in Driel at about 1400 and immediately met with the enemy fire, artillery, machine guns and single action fire (rifle)..

When asked how did I feel when I jumped, I was asked this by Prince Charles in Oosterbeek (post war celebration). I was very happy, I was glad to be jumping as we had trained for all these years. When I saw the light (tracers) I knew I was in action, and I was happy. When we jumped we found out that there was a considerable part of the brigade missing, first we did not know, later we found out, that while we were flying, the RAF ordered our planes to return as the weather was so bad that we should not jump. Only 41 aircraft heard this order and returned to England. We had a very good drop, the weather in Holland became clear and nice, the only trouble was that we did not know what had happened to 1/3 of our Brigade. The planes that did not turn back and did fly us to the DZ did not pick up this order from the RAF (as while we were in the air, we were subordinate to the RAF, and as any force would be, until we departed the aircraft). Our Pilots (those who continued) did not pick up the 'return' message.

Confusion at Arnhem ... when our gliders landed on the 19th (September) they came under heavy German fire, this area (LZ) was supposed to be safe (secured by the British, in fact the British had to move from the LZ as it became unobtainable, unfortunately this was not known to the Poles when they landed). Some Glidermen were taken prisoner right as they landed. Thus the Poles and Germans were in the same area, and British troops did fire on the Poles, thinking they were German, and then vice versa. These Poles were of the anti tank section (who fought along side of the British throughout the battle).

When we landed in Driel, we knew there was some sort of a ferry that was to take us across the river, as this was our orders, to join the British on the other side. We went to where this ferry was supposed to be, it was not there, something had happened to it. We had not boats of our own, so the general (Sosabowski) sent some Sappers (Engineers) down the river to try and find some boats, but they did not find any.

While we were waiting for boats (and we waited for some time), a Polish Officer (captain Zwloanski) swam across the river and told the General that the Division (1st Airborne) was in a very bad state and that General Urquhart requested that General Sosabowski should carry on with the crossing, and that he (Urquhart) would supply some boats and widen the perimeter (this area was where the British Airborne were surrounded) so as to accommodate the Poles with an area of space, as at the moment there was not enough room within the perimeter, as it was very small. All around this perimeter were the German. Urquhart was not really able to do this, the Division was in no position to do this, so by 4 o'clock in the morning General Sosabowski decided to go back to Driel, as we were maybe a mile or two from any cover and we were in the open.

In Driel orders were issued to make defensive positions. At about 7:30AM I was called by a radio operator, calling for Kuba, "Hello Kuba", this was Capt. Zwolanski, who had swam back to Urquhart (who was liaison with the 1st Division), Zwolanski was calling again and again as he had not heard from me. We then made our radio contact at 0730 September 22nd 1944. We both were being attacked in Driel and in Oosterbeek, radio contact was difficult to make as we were moving a lot and were fighting.

There were about 5 or 6 German tanks that were shooting at us properly and supported by an infantry battalion. We were in a problem because we hadn't any anti tank artillery with us (it was all on the other side of the river [with the British]). We had to wait until they came close enough so we could throw some mines at them. Then some of our men brought two British Armored cars with them who had come from the south, as part of XXX Corps advance guard. General went to them and asked for some support, as when asked by other subordinate Polish officers these British Recce troops declined, saying that they were not combat troops, but rather Reconnaissance troops and they did not want to open fire. So the General traveled on the famous ladies bicycle to ask them to fire on the German tanks. When they did fire, the Germans must have though that we had some anti tank artillery, so they withdrew (the Germans pulled back their armor). This attack stopped. From here we were always ready for them. At this time there was some movement in Nijmegen and they (Germans) relented and concerned themselves with the Americans and British XXX Corps.

We were trying to cross the river somehow, and some of our patrols went to our jumping area (DZ) and searched for some equipment that might help us. In some of this equipment they found some dingeys (two man rubber boats, meant for downed air crewman). We prepared for a night crossing, though the dingeys were for two people and there were no oars, so the crossing troops had to use spades as oars (paddles). On that night only about 56 people crossed the river, all of the dingeys were damaged by the German shooting at us while crossing. These men who crossed were divided up by the British and used immediately where ever there was a dangerous area (gaps in the defenses).

In the morning our chief of staff went south and asked for some boats, and was promised these boats. We prepared very seriously for this crossing as this was to be our last crossing and it was very important. We were supposed to get these boats about 10:30PM, instead they arrived about 00:50 the next day. Normal war time problems were happening. We started our crossing well after 0300AM.

As the dikes were slippery, some of the boats and amphibious vehicles slipped and went into the water and never came out, sinking in the water. These were the DWKS, sent by XXX Corps, not suited to steep inclines like that of the dikes of Holland.

We were divided into groups of 18 people to a boat, this is what we were told. But when the boats arrived the boats were for only 12 people. Our Brigade was marching to the boat, 6 men were left over, causing some commotion, as each group of 18 were planned to go as a group (radio teams, machine gun teams, those who worked together as a group). And all this time we were with the German firing at us (the river bank was under constant fire form the enemy). We had been under fire very strongly, particular this time was the heaviest. The ground felt as if it were moving up and down from the big German mortars, the explosions were very heavy.

The German burnt two large factories (which overlooked the river) so the river banks were very visible. There was no cover there for us (the Poles were completely exposed). At one time we had an Artillery observer (British) to direct the fire to support us. I went to show him where the four German machine guns were. That night we did get some artillery support (XXX Corps) but these guns were out of range (too far back) to be affective for us (the targets were maybe 11 miles and the guns could reach 10 miles) as the British Land Army did not manage to go as quickly as they were supposed to. We were having some casualties on the river bank. I saw the general running toward me and he said "Lorys, you have to stop all the movement and reorganize the men to 12 (allotment per boat). I went back to carry out the Generals order. At Dawn, the General stopped the crossing.

I have never been able to count how many Poles crossed the river, some say 200, 250, or 300. We came back to Driel and it was in ruin from the German artillery fire. We had a hospital in Driel and the hospital had been hit, and it was something terrible. People were being operated on and the doctors were working and now these men (the wounded) had to be hit on a second time, it was shocking! I only mention this as the whole house were plastered with red and white flags-showing that this was hospital and not fighting place and the German were shooting at it deliberately.

During this day we had a visit by a General Dempsey in Driel, he had had a talk with Urquhart. Urquhart was trying to stress how difficult things were on their side of the river. General Horrocks also came to General Sosabowski, who said to Horrocks that it was no use to make another small attempt like we were trying (to cross the river) and that there should be a bigger attempt further west and from behind the Germans with an entire Division. In Sosabowskis opinion the German are not very strong and will not be able to hold off an entire Division (attacking form behind their lines). If they were strong enough, then they would have over run the British Airborne by now (if not earlier). They, the German would have been able to overrun the British, as the British were out of food, ammunition, they were out of everything ... but the German could still not overrun them because they themselves were not in strong enough force to do so. Sosabowski considered that a strong enough force would wipe out this German force and the objective (Arnhem) would be achieved. Horrocks listened, and then told Sosabowski to come to a conference this morning at the 43rd Division's HQ. By this time it had been 9 days since the British had been holding on in Arnhem, originally being told that it would be 48 hours, the most 72 until relieved.

Chapter Fourteen: 'Tobie Ojczyzno' - The 1st Polish Independent Parachute Brigade

The conference was set up much like a court marshall, with one chair in front of a large table (where all others sat) which was for the general (Sosabowski) on its own. Horrocks, and Browning were there among others, Sosabowski and Dyrda (Dyrda was often with the general, and was skilled diplomatically in dealing with those in the general staff) – the appearance was shocking to Dyrda. Commanding Officer General Thomas of the 43rd Div. was to take over the crossing and the Poles were to be subordinated to him. Thomas delivered his plan, which was for the Dorsets battalion (43rd Div.) to cross just west of the Poles position. Sosabowski brought fourth his question, saying that this plan is due to fail (which it did), and that we should cross further west in a bigger force and hit the German from behind (as he had discussed with Horocks that morning. Thompson said, "The plan is as I said", Sosabowski tried to say something, and then Horrocks said "General Sosabowski if you are not willing to execute our orders we will find another officer to replace you." Sosabowski said, "There was no question that I was not going to follow your orders." Then all left and went about their business.

That night the boats for the Dorsets never arrived. A request came from the Dorsets asking for the Polish boats, General Sosabowski agreed that they may have the boats first, as the plan was for the Dorsets to cross first. By about 0200 the Dorsets managed to get across about 300 men (few reached the British Airborne, most were killed and or taken prisoner as they reached the other side). The German were shooting at them as they were shooting the Poles the previous night. Because we had no boats, we did not cross. This was the last attempt to rescue the British Division (Airborne).

The next day orders came for the British across the river to withdraw. We received our orders, that we are to stay there on the riverbank and show the British (evacuees) the way to tea wagons and then send them off to Nijmegen, where truck were waiting for them (taking them to awaiting aircraft, which would take them to England). For this withdrawal there were some big boats (Canadian motor boats, and some with oars), about 1,400 British (Poles included) those who had crossed the river and not been killed or captured, and the surviving members of the anti tank section who were lucky enough to make it. Troops were evacuated until it was daylight, and then the operation had to quit due to German fire. The Poles were the far rearguard, came to the river and had to jump in the river and try to swim as there were no more boats, some drowned and/or were shot while swimming. These were the British, Canadian can loans (Officers and Specialists on loan from Canada), and Poles. Things became desperate, knowing that no more boats would come. Some swam out to the last boats as they left, others made it on their own to the other side, and some simply surrendered as the German had become aware that something was happening and surrounded the area and subjected it to heavy fire. White flags came out and the Germans came down and gathered up those who were left. Polish Chaplain Misiuda jumped into the river, attempting to swim across, and his body was never found. And that was the end of the Battle of Arnhem. (*These notes from the Hartenstein Museum Oosterbeek: For the withdrawal there were some Canadian motor boats and about 1,400 British boats that were used to ferry the troops across the river. There were a few Canadian motor boats that ended up in the wrong area place to ferry the troops. The number of soldiers that crossed the river that night was 2,398. This is broken down into the following: 2,163 of the 1st AB Division, 160 Polish, 75 Dorsets, a select number of the RAF (Pilots and Liaison Officers), 1 Dutch Jew, and 1 German POW. Also of note is that Polish Chaplain Misiuda was killed on the way to the river and had a field grave on the northern side of the river. This is mentioned In the book 'Pole's Apart'.)

Dyrda told me that after the conference with General Thomas, that General Browning took Sosabowski to dinner. There Sosabowski asked Browning, "Why we are planning such a small crossing instead of a bigger one, the size of a division further west?" Browning replied that they hadn't enough boats for such a crossing. Sosabowski then replied, "But General Browning there are so many cars here and so many trucks and there are no boats? How is this possible?" Browning was not happy, thus Sosabowski put another nail in his own coffin, politically (Browning had such a large and extravagant 'train' [HQ allotment] in Nijmegen that extra gliders were used to transport these, gliders that could have been used to carry more troops and or equipment's that were badly needed).

From Arnhem, we were sent to defend airfields west of Nijmegen. I was called to go to London to go to America to take a Staff course.

There were several of us, Airmen, Tankmen, Paratroopers, all Polish to take part in this American general staff college course (Ft. Leavenworth, Kansas). I learned very much in this school. I remember that in the American Army a star is the rank of a general, we Poles used stars from the rank of 2nd Lt. To Captain (3 stars) – everybody was saluting us, higher ranking officers included. After the college I went to Ft. Benning, to get experiences to see how they were getting troops ready to go to the front (the UK and Europe). I remember seeing some beginning jumpers in the trees (hanging by their chutes, caught in the trees), making them look like Christmas trees, as the Americans dropped these men in strong wind. I jumped in one of these exercises and I remember seeing this.

I was drinking with some of my American friends (at Ft. Benning) and I told them of how we had once been dropped up to five miles from our target. The Americans said, "Jan, when we drop you, you will be able to land on an x!" I did not believe this and bet them a bottle of whiskey. I jumped and as I was coming down I thought first yes the whiskey is mine, but as I came down I came closer and closer to the x, and finally landed on the x, so I was out a bottle of whiskey.

In the UK, the Brigade received huge replacements and was retrained and ready for action in May of 1945. We got ready to make our last Brigade jump (the entire brigade was to do a practice jump), and we were ready on the airfield, we were then recalled and orders to go to the continent to take over from the Czechs (who had gone back to Czechoslovakia) who were guarding Dunkirk. We were on boats and the day was May 8th – the day the war in Europe ended. We then spent 2 years occupation duty in Germany. Gradually we were withdrawn to England and the Brigade was disbanded in 1947.

For King and Country: British Airborne Uniforms, Insignia & Equipment in World War II

Above: Here are two bicycle mounted gliderborne troops – glider troops, as they are equipped with the Bergen Rucksacks, as if they were paratroops they would simply have the web haversacks. On the left is an officer with the 1st series, 2nd pattern folding bike – this is identified as such because of the khaki paint and very low frame number. We know he is an officer as he is wearing a tie, and he is wearing a 2nd pattern Denison which has been converted to full zip, and the experimental body armour as issued to the Poles for the Arnhem operation. Just visible is the rim underneath the camouflaged helmet net – this is a fibre rim helmet which was commonly worn by the Poles. The net is split colored, half chocolate and half olive green. These nets seem to have been issued just prior to Normandy and were worn until the end of the war. The canvas bag attached to the bike is the 1st pattern 'oval' bike bag which is attached to the bike via P-37 web straps. On the right is the more common 2nd series, 2nd pattern folding bicycle with BSA repair kit strapped to the frame and the pack frame (Bergen) strapped to the front of the bike. The Denison is a 1st pattern as denoted by the haphazard manner in which the camouflage paint was applied. 1st patterns commonly have this sand or yellowish base color to them, and so stand out from the screen printed pattern of the 2nd pattern smocks. A 2nd pattern Sten bandoleer is worn with a 3rd pattern web chin strap jump helmet.

Right: Captain Jan Lorys is seen here wearing his Service Dress uniform which was tailor made for him. The buttons were of white metal with the Polish eagle embossed. This photo was taken after the war as he wears the jump wing with combat wreath. Visible are his rank stars on his shoulder straps, three on each side for the rank of Captain. (Photo courtesy of Jan Lorys)

176

Chapter Fourteen: 'Tobie Ojczyzno' - The 1st Polish Independent Parachute Brigade

A Sunday service as visible to the far left above the bell tent, a Chaplain in mid-ceremony. The troops in the rear are all wearing the MK II helmet as worn by line units. This dates this photograph to between 1940 and 1942.

Poles on a march somewhere in England, possibly back from a day's exercises. Note the manner in which the MK II Stens are carried – the Poles commonly carried both their rifles and Stens in this fashion, to the front and slung over the neck and left shoulder. All troops wear the fibre rim helmet.

The Poles at inspection. Note the No.1 rifles and MK II Sten as worn by the third man in line who is being inspected. The berets worn are of the 1st pattern. Also note that each man holds a booklet in his hands that is also being inspected and notate that this soldier passed or did not pass inspection, and that all his kit was in-order. Look closely for the golden eagles painted on the front of the helmets.

An early shot of the men on maneuvers. Note the No.1 rifles and the modified Irvin Jackets. This is a signals platoon with eight No.38 sets. The poles are actually antennas which could be carried in a special P-37 web carrier. Here the 2nd pattern jump helmets are worn with eagles stenciled on their fronts, as well as 1st pattern berets with 1st pattern cloth beret badges. The toe plates and heel cleats are very noticeable on the signaler first row, on the far left.

The troops at a presentation. Note that both the Polish and British flags are flown side by side. By the look of the helmets, most seem to be the 2nd all steel 1943 pattern, with one 1st pattern fibre rim (first man on far right, second row).

General Sosabowski with British general staff officers and Polish officers discussing the 75mm howitzers. Some are broken down as the Poles, as well as the British, did not drop these guns, but rather brought them in by glider.

Left: A group of paratroops having a meal. The food is warmed up on a British 'Tommy Cooker' and then served. To the right is an NCO tape which is on the shoulder strap of the soldier preparing the meal. His rank is of Lance Corporal, being of one piece of NCO tape. Two tapes side by side would be a full Corporal, and three a Lance Sergeant. The white between his blouse and trousers is the trouser size label, and note that he is wearing the parachutist trousers. To the L.Cpl.'s right is a 2nd Lieutenant, as he wears but one rank star beneath his cloth 1st pattern beret badge on his beret. Also note that he is wearing collared shirt and tie. The Lieutenant's smock is a 2nd pattern, and the trooper to his right wears a 1st pattern smock and a metal badge on his beret. Of interest note the pannier conversion done to the jeep's rear section. Right: At the bottom left the battledress blouse is buttoned to the battledress trousers. The dark chocolate buttons found on the L.Cpl.'s blouse make it an earlier dated Austerity pattern, probably from 1942.

Somewhere, outside of a little village in England these Poles stopped to take a breather and a quick photo opportunity. As all seem to be wearing Denisons in place of the modified Irvin Jackets, this can be dated 1943 to early 1944. It is difficult to make out the make of rifle, though the soldier standing beneath the bridge, second from the right does looks like he holds a No.4 rifle.

Chapter Fourteen: 'Tobie Ojczyzno' - The 1st Polish Independent Parachute Brigade

A group on maneuvers in England prior to Market Garden. On the bonnet is an airborne sleeping bag – note its protective lined bottom. Also note the lack of windscreen, and the bonnet rack just under the sleeping bag which was for the storage of ammunition. The trooper standing wears a P-37 sleeve water bottle carrier.

A lieutenant gives instructions to the trooper which he will 'run' to HQ. Note the officer's airborne folding map case. The brass ring attached to a shoulder strap and was worn to the front. The berets seem to be the 2nd pattern which were dark gray, as opposed to the lighter poster gray of the 1st patterns. The officer wears a 1st pattern beret badge with his rank star underneath. The Signaler sits patiently awaiting orders. The headphones are standard issue. Of interest the driver wears the metal parachute collar insignia which are backed on gray felt and edged in yellow for paratroops.

For King and Country: British Airborne Uniforms, Insignia & Equipment in World War II

A group of Poles on the LZ awaiting radio contact with 1st Airborne as they were not at the LZ as planned. As a note 1st Airborne had to pull back and the area between them and the LZ was overrun by the Germans. Note that the jeep has that stripped look – this was done to ease the loading procedure, as well as to lighten the load. This jeep features a detachable steering wheel, a reinforced bumper and a tandem towing hook. Also note that the spade is attached to the bumper, and on the hood is the airborne sleeping bag.

Here a wicker supply pannier is retrieved from the brush where it landed. The colored triangles denote the type of contents found inside. Note the metal collar insignia worn by the trooper on the right. He is also lifting the pannier by its wicker carry handles, which are found on each corner. As can be seen these panniers are quite big, which explains why special modifications were made to the jeeps for them to carry the panniers.

The wounded were transported via jeep, and each medical jeep was specially fitted with stretcher racks (variations of this will be covered in a future volume). Note the airborne modification of holding the Jerricans behind the seats. The wounded trooper has been dressed with a shell dressing.

A Polish medical orderly. He carries two, one on each side, P-37 shell dressing bag slung over his shoulders, and holds the 1st pattern airborne folding stretcher. The difference between the 1st and 2nd pattern is that the 1st had balled ended legs as seen here. This stretcher could be strapped to the man's leg and released like a rifle or Bren valise prior to landing.

Here members of the Brigade are eating their canned 'drippings', and assorted rations. Sitting against the wall, in the group of four, second from the right is an NCO – notice his edged shoulder boards. It is difficult to tell what rank of sergeant that he holds, he is at least a Sergeant to Staff Sergeant. The man eating on the far right, closest to the frame is a Corporal. Most smocks and equipment have been removed, making it look like the end of an operation – possibly after the Brigade withdrew from Driel. Of interest are the 1st and 2nd pattern berets worn side by side – the man on the far left wears a 1st pattern, as it is lighter in color, the three next to him are darker in color, therefore 2nd pattern berets.

Chapter Fourteen: 'Tobie Ojczyzno' - The 1st Polish Independent Parachute Brigade

A group on maneuvers in England prior to Market Garden. On the bonnet is an airborne sleeping bag – note its protective lined bottom. Also note the lack of windscreen, and the bonnet rack just under the sleeping bag which was for the storage of ammunition. The trooper standing wears a P-37 sleeve water bottle carrier.

A lieutenant gives instructions to the trooper which he will 'run' to HQ. Note the officer's airborne folding map case. The brass ring attached to a shoulder strap and was worn to the front. The berets seem to be the 2nd pattern which were dark gray, as opposed to the lighter poster gray of the 1st patterns. The officer wears a 1st pattern beret badge with his rank star underneath. The Signaler sits patiently awaiting orders. The headphones are standard issue. Of interest the driver wears the metal parachute collar insignia which are backed on gray felt and edged in yellow for paratroops.

For King and Country: British Airborne Uniforms, Insignia & Equipment in World War II

Two troopers prior to the Arnhem jump, and behind them the C-47 Dakota transport. As by their rank stars, found upon their 1st pattern berets, on the right is a Captain, and on the left a 2nd Lieutenant. Both officers wear the parachutist trousers. The Captain has a pistol lanyard attached to his side arm and hung from around his neck.

Here is a group of Engineers. Note that three men are wearing the V neck pull over, and the 2nd lieutenant has the airborne map case with its strap around his neck. The older officer is wearing a leather jerkin. All wear the 1st pattern beret. Of interest is the airborne cart, which weighed some forty pounds, and was an all purpose cart which was delivered by glider. Underneath the cart are spades.

Chapter Fourteen: 'Tobie Ojczyzno' - The 1st Polish Independent Parachute Brigade

Here is a Lance Corporal at attention for the morning flag raising. Note the golden stenciled eagle on his helmet, and the all cloth 1st pattern collar kites, edged in yellow for paratroops. He wears the metal jump badge without combat wreath. His ribbons are, from the left: Service Cross, the War Medal (British), and the Defense Medal (British). His blouse is a Battledress serge 1940 pattern, as by the size label.

The author as a 2nd Lieutenant. Note the single rank star on the shoulder straps, and on the first pattern beret with 1st pattern cloth badge. The jump wing is without combat wreath (remember this wreath only came out after Arnhem, therefore any photos that feature wings with the wreath are very near to, or on the line of post-war). The collar kites are un-edged, as was a post-war practice. The blouse is an Austerity pattern dated 1945.

Above: A group of troopers rush to open and empty a just delivered container, dropped by the RASC Air Dispatch. Note the colored identification band on the container. Of interest are the FS knives carried in their special pockets of the parachutist trousers. Note the difference in color on the Officer's helmet net, half chocolate and half olive green.

Right: The container is opened and its contents unpacked. Notic that its contents have been carefully wrapped in heavy felt and blankets to prevent damage upon landing. At the top of the container is a series of D rings, these are to attach the parachute bag to – there are four D rings in total. The chute and bag sit inside this 'cupped' opening. Just visible is the single rank star on the officer's shoulder strap. This container was found as is, painted orange, when it was dug up in the Arnhem area, and was used later in the film, "A Bridge Too Far."

181

For King and Country: British Airborne Uniforms, Insignia & Equipment in World War II

A group of Poles on the LZ awaiting radio contact with 1st Airborne as they were not at the LZ as planned. As a note 1st Airborne had to pull back and the area between them and the LZ was overrun by the Germans. Note that the jeep has that stripped look – this was done to ease the loading procedure, as well as to lighten the load. This jeep features a detachable steering wheel, a reinforced bumper and a tandem towing hook. Also note that the spade is attached to the bumper, and on the hood is the airborne sleeping bag.

Here a wicker supply pannier is retrieved from the brush where it landed. The colored triangles denote the type of contents found inside. Note the metal collar insignia worn by the trooper on the right. He is also lifting the pannier by its wicker carry handles, which are found on each corner. As can be seen these panniers are quite big, which explains why special modifications were made to the jeeps for them to carry the panniers.

A Polish medical orderly. He carries two, one on each side, P-37 shell dressing bag slung over his shoulders, and holds the 1st pattern airborne folding stretcher. The difference between the 1st and 2nd pattern is that the 1st had balled ended legs as seen here. This stretcher could be strapped to the man's leg and released like a rifle or Bren valise prior to landing.

The wounded were transported via jeep, and each medical jeep was specially fitted with stretcher racks (variations of this will be covered in a future volume). Note the airborne modification of holding the Jerricans behind the seats. The wounded trooper has been dressed with a shell dressing.

Here members of the Brigade are eating their canned 'drippings', and assorted rations. Sitting against the wall, in the group of four, second from the right is an NCO – notice his edged shoulder boards. It is difficult to tell what rank of sergeant that he holds, he is at least a Sergeant to Staff Sergeant. The man eating on the far right, closest to the frame is a Corporal. Most smocks and equipment have been removed, making it look like the end of an operation – possibly after the Brigade withdrew from Driel. Of interest are the 1st and 2nd pattern berets worn side by side – the man on the far left wears a 1st pattern, as it is lighter in color, the three next to him are darker in color, therefore 2nd pattern berets.

Chapter Fourteen: 'Tobie Ojczyzno' - The 1st Polish Independent Parachute Brigade

Here three officers pose for a photograph. From the left, a Major with two tapes and one star upon his shoulder straps and beret, in the middle a Captain with three stars under his beret badge, and on the right a 1st Lieutenant with two stars upon his shoulder straps and underneath his beret badge. All wear 1st pattern berets and the early issue battledress serge blouses and trousers. It looks as if the Captain is wearing a 1st pattern smock as noted by the long solid patch of color down his left sleeve – a 2nd pattern's scheme did not spread in this fashion.

It is difficult to date and place this shot. Shown here are members of the Brigade and 1st Airborne. This may be part of the seaborne element of the 1st Division prior to the Arnhem jump. The two on the far left are British as by their maroon berets. The man on the far left is a Captain as by the three pips on his Denison shoulder strap. Also note how he wears his pistol lanyard around his arm rather than around the neck. Next to them is a Polish Captain, as he has three stars upon his beret. In the middle front is a 2nd Lieutenant wearing parachutist trousers with an airborne map case, and the larger revolver holster as he is armed with a .45 (note the .45 magazine pouch just above the holster). The captain also seems to be armed like the 2nd Lieutenant. Next to the 2nd Lieutenant is a British officer. In the second row is another Polish Captain. Most seem to be wearing the standard battledress trousers.

Battledress

The Polish Brigade wore all patterns of Battledress as well as the Specialist Parachutist Trousers throughout the war. They were heavily modified by the Poles. Shoulder straps, collars, and bodies of the blouses and trousers were specially tailored, not only for officers, but other ranks as well. The Poles took great pride in their appearance, going to great lengths to assure a smart look.

Berets and beret badges

Only upon a third submission by Sosabowski, on April 28, 1942 was the first Polish airborne beret approved by the commander-in-chief. A first submission was a FS (field service) cap with blue top. A second was a khaki beret, with 28 string tassel representing the 28 shroud lines of the parachute.

Made of a 'poster gray' color, it was issued and worn with a special cloth cap badge of a white eagle on a poster gray back ground for other ranks. And silver wire bullion for officers on the same poster gray backing. Prior to this, the British Field Service cap was worn with a standard metal eagle.

As materials were in short supply, the British War Office and Polish authorities took steps to manufacture a new beret. In mid-1942, the British were hoping the Poles would use the RAF blue worsted wool since it was available and would save money for the war effort. In July 1943 Sosabowski came up with a dark gray beret of a 'conventional pattern' meaning, that it was of a one piece construction. The first pattern was of a three piece construction. The second pattern was approved, and then issued in July 1943 and worn until disbandment. The second pattern dark gray beret was to be worn with the white metal cap eagles and not the cloth eagles as made specially for the first pattern berets. As with anything, both patterns can be seen in use side by side throughout the war. Officers were known to wear either the silver wire or cloth 'other ranks' cap/beret badges.

Plastic economy cap eagles were worn in very limited number. I have seen but one example of this. They were introduced on January 29, 1944, and were discontinued on June 18, 1945. There were some 136,500 of these plastic eagles manufactured.

Here is a battledress serge 1940 pattern. The jump badge on this blouse had its tail sewn down as seen – this was not regulation. The NCO tape, a single as shown, denotes the rank of Lance Corporal. Note the 1st pattern collar kites.

A1st pattern beret with 1st pattern beret badge. Beneath is a Sergeant's beret chevron, and 1st pattern collar kites. Note how the yellow edging is attached to the gray kite. To the right is the jump badge with its screw plate. The badge has a screw attached to the back of it, a small hole is made in the battledress blouse and the badge stuck through with the plate acting as a washer, securing the badge. Note how the beret is made in three pieces as opposed to the British one piece construction. This is why the British wanted a simpler, less costly variant produced for the Poles – thus the one piece 2nd pattern.

A group taking a breather while on maneuvers. Note that all wear the modified Irvin Jacket. Eagles are found stenciled on the fibre rim helmets, and 1st pattern berets are worn. This photo looks to be from 1942, as the men are armed with Mk III Stens and rifles. Also seen braced against the stone wall is a Bren LMG. The sixth man wearing a beret is a signaler, as he has a No.38 set. Of interest is the helmet lining of the man second from the left – here can be seen the cotton tapes which allow the helmet to be worn comfortably and/or adjusted to the depth of the wearer's crown.

Right: From the left: 1st pattern cloth beret badge, plastic economy badge, and the metal badge as worn on the 2nd pattern beret. The plastic was not limited to airborne, rather more so worn by the line and armored troops. There are many variations of the metal badges, larger, smaller, brass and such.

Chapter Fourteen: 'Tobie Ojczyzno' - The 1st Polish Independent Parachute Brigade

Jump helmets

The Poles were issued the Bungey helmets for training which had the large golden eagles stenciled upon the front center. Later the 2nd pattern fibre rim helmets with leather chin straps were issued. Some Poles replaced the leather chin strap with the 1943 web chin strap on the 1942 helmet. By the Arnhem Battle September 1944 50% of those worn were the late war pattern helmet with web chin strap. A golden eagle was stenciled on each shell. This not only served as an identification marking for the brigade but also the division. Special unit flashes were painted on each side of each shell for the Arnhem operation yet it seems to have been done predominately only on the left side of the shell. This made identification easier within the brigade. It is very difficult to see this in photographs as most helmets were covered with the camouflage net and scrim but the markings are there.

The stamping of the Polish jump helmet. Visible on the top is the manufacturer, then size, and then date (1944), and to the right is the broad arrow – the 4 PLT stands for 4th Platoon. This helmet supposedly belonged to an Englishman who was attached to the Poles, and would explain the British marking 4 PLT – his name was to the left of the 4PLT. Notice how the draw string of the net is drawn tight and rolled up underneath the edge of the helmet shell.

Seen here is the common eagle stenciling, always in yellow for the airborne, on this 3rd pattern web chin strap helmet. White eagles were painted on after the war and are commonly seen worn for occupation duty – these white eagles were much larger. Note the light olive wash applied to the helmet to reduce glare – this same wash is often found on fibre rim helmets. This helmet does not have the recognition flash painted on its side, as most did for Arnhem.

An unusual 3rd pattern helmet. All eagle stencils the author has seen are like that found in the previous photo. The eagle on this helmet is very crude, yet it is interesting as the shape of the eagle is elongated like some of the metal beret badge variants. The helmet rests on the back plate of body armour. (The DeTrez collection)

The inside of this unusually painted 3rd pattern helmet. Of interest is how the two rear pieces of leather are glued onto the rubber pad sections. There are three sections of rubber – where each buckle is attached, the rubber ends and then another piece begins, thus three pieces of foam rubber in the construction.

Jump Smocks

The first smocks that the Poles are known to have used are what appear to be modified Irvin Flight Crew jackets. It is not known whether the Irvin Company (the same Irvin Co. that manufactured parachutes for the Airborne forces) made these as a custom run for the Polish Brigade, or that the Poles themselves simply made the modifications. The Irvin Flight Crew jacket looks very much like a 1st Pattern Step-in denim One Piece, yet in many of the photos shown here this garment is indeed quite different form the One Piece Step-in smock. As the Poles were low on the "totem pole" to be issued with new equipoement they had top make the best with what they were given. These modified Irvin Jackets were worn up until the Polish Brigade came under British command. Once this took place the Brigade was completely re-outfitted with everything available to their British counterparts. This included the issue of 1st Pattern Denison Smocks.

Insignia

Jump Badge

Designed by Marian Walentynowicz, who took the design from the sleeve of the English published book, *The Earth is Gathering Ashes*. On September 23, 1941 the first combat exercises that the Polish airborne took part in were in Kincraig, Scotland. On that day, the 4th Polish Rifles Brigade were redesigned the 1st Polish Independent Parachute Brigade. On this same day, the first 'Parachute Badges' were awarded. Metal in construction, each badge was secured to the BD blouse by a screw and washer. The badges were only worn on the Battledress blouse. On the back of each badge was the dedication, 'Tobie Ojczyzno' ('For You, My Country'), and a serial number. 6,536 badges were manufactured, thus each badge can be traced to the awardee (see 'Materialy' by Capt. Jan Lorys 1993, available through the Sikorski Institute, in London).

The badges were awarded to those who successfully completed parachute training in a Polish unit. The Poles trained not only themselves, but others, as room elsewhere in the UK was very limited. 238 badges were awarded to the French, 172 to the Norwegians, 46 to the British, 4 to the Belgians, 4 to the Dutch, 2 to the Czechs and 3 to the Americans. On November 20 1944, changes were ordered, and from that day the eagle was to have a crown of laurel in his claws to commemorate those who had taken part in the Battle of Arnhem two months previous. This being the difference between the standard jump badge (wing) and combat jump badge.

There was a glider and combat glider badge, yet these were not made or issued until February 1945. At this time the Poles were not fighting but assigned to 'guard duties' of air fields and prisoners of war.

Shoulder Titles

The Polish Brigade wore both embroidered and printed shoulder titles, though the embroidered were prevalent amongst the troops. Officers could wear bullion wire titles but this was done more so on their Service Dress.

Collar Patches

Designed to be worn on Battledress and great coats. For the BD, there are three variants, a dove gray worsted 'kite' edged in yellow, featuring a white parachute, dark gray 'kite' edged in yellow, featuring a white parachute, and a dark gray 'kite' edged in yellow yet with a metal parachute and paratrooper.

These are the basic collar badges for the airborne infantry. In August 1942 Officers were authorized to wear kites in the color of their branch of service. Other ranks also wore the kites and metal chutes, though they were not officially authorized.

The Poles specially tailored their battledress as seen here – the collars almost seem as if they have been stretched. The collars were commonly re-cut and thinned and the collar points enlarged. The man at right has added the metal or plastic Polish buttons which feature the Polish eagle. Also note how he has added pleats to his Austerity pattern battledress blouse. The little guy in the middle, who is probably the toughest of the lot, wears a modified Austerity pattern blouse, and on the left an earlier battledress serge blouse is worn. All troopers wear the jump badge with combat wreath, thus this was taken after Arnhem, probably in 1945.

This looks like an occupation duty shot, taken in Germany. Note the jump badge with combat wreath. This soldier wears a tie, yet no officer's stars are seen on his shoulder straps, therefore this must be taken in very late war or afterwards. He wears a modified battledress serge blouse, as his lapels have been lined with extra pieces of serge.

There was a post-war worsted kite which was unedged. This was seen worn on war dated blouses. These seem to be of a special issue for the BAOR (British Army of Occupation of the Rhine in 1945). I have seen two blouses where the original kites had been replaced by the post-war kite badges.

Chapter Fourteen: 'Tobie Ojczyzno' - The 1st Polish Independent Parachute Brigade

Badges of Rank

Officers wore metal stars much like the British wore pips. Some are screw back others have small teeth that are stuck through the shoulder strap and then folded over to secure it. Stars were placed straight across the shoulder strap of the Battledress. Rank went from one star for a 2nd Lt., to two stars for a 1st Lt., to three stars for a Capt. From here up a special bar was used to distinguish a major, Lt. Col. and up. These were also worn on the beret.

Other ranks wore a special NCO tape looking much like German tresse. It was silver in color with amaranth edging on each side. One tape loop for LCpl., two loops for Cpl., three for Lance Sergeant. From there it went to a series of inverted 'V's made of the same NCO tape with the edges of the shoulder strap 'lined' with this NCO tape as well. Rank was not worn on the Denison by either officers or other ranks. NCOs wore metal V's below the eagle on their berets. One V for Cpl., two for Lance Sergeant.

A 1st pattern collar kite, edged in yellow for paratroops. Note that this is a Canadian battledress blouse.

An officers Service Dress tunic and SD collar badges which were made only for the SD tunic. The parachute is white metal – these were also worn on the battledress, yet backed by the kite shaped gray badges and then edged. On the shoulder strap is an economy plastic Polish SD button. As seen earlier, some soldiers were known to sew these to their battledress, which was not regulation. Note the officer's quality jump badge.

Three variants of shoulder titles that were worn by the Brigade, from the top: embroidered, printed economy, and officer's silver wire.

Here you see some very early 1st pattern kites edged in yellow, again, for paratroops. As can be seen the parachute shrouds are 'lined and couched'. Beneath is a Sergeant's beret chevron.

The officer's rank star upon the shoulder strap. Note the officer's silver wire shoulder title

187

Epilogue

Words of wisdom for the treasure hunter

Some items featured in this book are next to unobtainable. I have spent many years digging through bins and boxes looking for some of these goodies and have found some. When I did I thought it was at the very least equal to the discovery of the Titanic. Now and then you'll get lucky and find a treasure and when you do, you should consider yourself very fortunate.

Condition. What many collectors fail to remember is that, there was a war. It was not a fashion show and a lot of these pieces collectors want went through it. They were not all sitting in a warehouse wrapped in protective coverings. Yes, you will find some pieces that are in truly unused 'never seen the light of day' condition yet there are many that have actually been in combat and are tatty and torn. Try to keep in mind that it is indeed military artifacts that you collect.

Sizes and prices. Prices per piece will differ according to size. The bigger the size the more expensive it will be. Collectors are infamous for wanting it all mint, unissued, war dated and in their size with their modern day girth. What they fail to understand or acknowledge is that people, especially the English were smaller in the 1940s. An average size was 5'5" breast 34" waist 30". Larger sizes were simply not made in any great number. Keep this in mind next time you say, "I want it mint, war dated and in an extra large size." Good luck and good hunting.

Bibliography

Cholewczynski, George F., *Poles Apart,* Sharpedon, 1993.

Davis, Brian L., *British Army Uniforms & Insignia Of World War Two,* Arms and Armour Press, 1992.

Duyts, W. J.M., Groeneweg, A., *The Harvest of Ten Years,* Airborne Museum Hartenstein Oosterbeek, 1998.

Englert, Juliusz L., Barbarski, I Krzysztof, *General Sosabowski,* 1996.

Harclercode, Peter, *Go To It Illustrated History of the 6th Airborne Division,* Bloomsbury, 1990.

Hoare, Tom, *Rouge Diablos One Man's War,* Avon Books, 1995.

Hobbs, Brien, *Military Illustrated,* No.26, July 1990.

His Majesty's Stationary Officer, *By Air To Battle,* 1945.

Lorys, Jan. *Materialy*, Sikorski Institute London, 1993.

Meel, van Rob, Bann, Monica, *British Airborne Jeeps 1942-1945,* Groucho Publishing 1997.

Middlebrook, Martin, *Arnhem Battle 1944,* 1995.

Saunders, Hilary St. George, *The Red Beret*, London Michael Joseph, 1949.

Skennerton, Ian, *British Small Arms of World War 2,* England, Greenhill books 1988.

Addendum

Late Additions

Variations of the 1st Pilot's brevet. Both were worn on the BD and Denison.

The printed variant of the SOUTH STAFFORD title.

The unofficial title S. STAFFORDS (gold on cherry). To the author's knowledge a printed variant was not made.

A nice Glider Pilot's Austerity blouse with the rank of Staff Sergeant. Note the brass crowns which are secured by poking through the serge wool and are held in place with a brass pin. Printed shoulder titles and Pegasus flashes are worn. Of interest is that AIRBORNE strips are not worn in conjunction as commonly seen.

Addendum: Late Additions

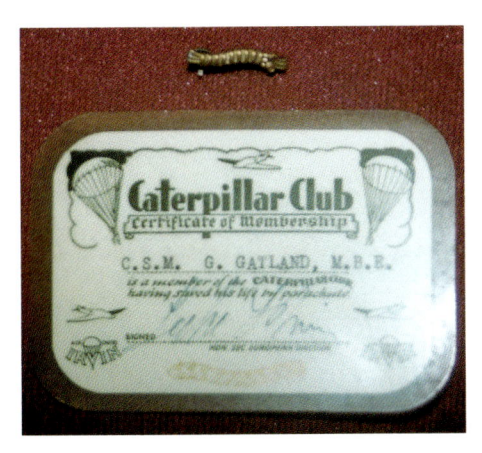

The Caterpillar pin and Club membership certificate. Note that Irvin (Irvin Parachute) presented the card.

Right: Shown here is a 3" mortar, mortar pit and crew. Note the leather tube cup which kept dirt from inside of the barrel.

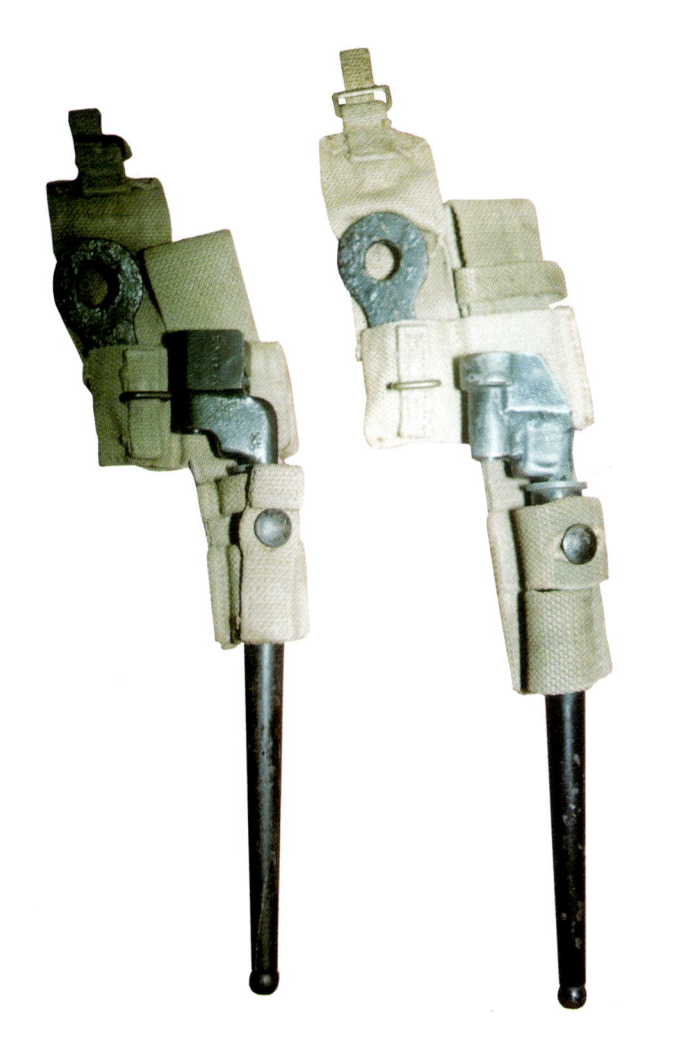

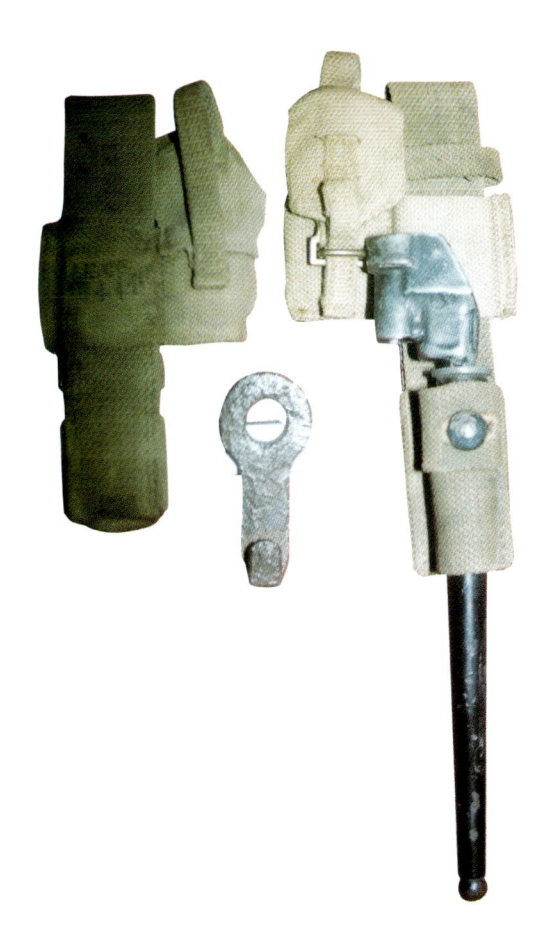

Left: Two Sten retaining plates that were found in the Arnhem area – inside of a rotted web Sten frogs – in the actual special pockets. Neither of the frogs survived the years and disintegrated upon excavation. Note the variation of the frogs shown. Right: Another look at the retaining plate. Note the loop on the left frog – this is how it fit over the waist belt and then through the frog loop which is to the left of that.

For King and Country: British Airborne Uniforms, Insignia & Equipment in World War II

Shown here are the nozzle, grip, and trigger assembly – hose is to the left.

A back view of the No.2 MkII Marsden "Lifebuoy" Flame Thrower. To the lower right is the hose (in black) – also note the web shoulder straps. This sapper is seated on a 6-pound anti-tank gun. The No.2 MkII was known to be unreliable as it depended on an ignition battery that, when wet, would not ignite. 7,500 No.2 MkIIs were produced in time for the D-Day invasion. Production was discontinued by July 1944, yet it saw service until the end of the war. The No.2 MkII weighted 64 pounds (with fuel), carried 4 Imperial Gallons of fuel, its range was 30-40 yards, and fired a 10 second continuous stream.

A close-up of the hose, and tank valve.